Fermenting
FOR
DUMMIES®
A Wiley Brand

by Marni Wasserman, CN
and Amy Jeanroy

Fermenting For Dummies®

Published by: **John Wiley & Sons, Inc.,** 111 River Street, Hoboken, NJ 07030-5774, www.wiley.com

Copyright © 2013 by John Wiley & Sons, Inc., Hoboken, New Jersey

Published simultaneously in Canada

No part of this publication may be reproduced, stored in a retrieval system or transmitted in any form or by any means, electronic, mechanical, photocopying, recording, scanning or otherwise, except as permitted under Sections 107 or 108 of the 1976 United States Copyright Act, without the prior written permission of the Publisher. Requests to the Publisher for permission should be addressed to the Permissions Department, John Wiley & Sons, Inc., 111 River Street, Hoboken, NJ 07030, (201) 748-6011, fax (201) 748-6008, or online at http://www.wiley.com/go/permissions.

Trademarks: Wiley, For Dummies, the Dummies Man logo, Dummies.com, Making Everything Easier, and related trade dress are trademarks or registered trademarks of John Wiley & Sons, Inc., and may not be used without written permission. All other trademarks are the property of their respective owners. John Wiley & Sons, Inc., is not associated with any product or vendor mentioned in this book.

For general information on our other products and services, please contact our Customer Care Department within the U.S. at 877-762-2974, outside the U.S. at 317-572-3993, or fax 317-572-4002. For technical support, please visit www.wiley.com/techsupport.

Wiley publishes in a variety of print and electronic formats and by print-on-demand. Some material included with standard print versions of this book may not be included in e-books or in print-on-demand. If this book refers to media such as a CD or DVD that is not included in the version you purchased, you may download this material at http://booksupport.wiley.com. For more information about Wiley products, visit www.wiley.com.

Library of Congress Control Number: 2013946294

ISBN 978-1-118-61568-3 (pbk); ISBN 978-1-118-80458-2 (ebk); ISBN 978-1-118-80469-8 (ebk)

Manufactured in the United States of America

10 9 8 7 6 5 4 3 2 1

Contents at a Glance

Recipes at a Glance

Table of Contents

Part V: Beer, Wine, and Other Beverages 203

Chapter 14: Healing Beverages 205

Chapter 15: Making Wine from Water and Fruit 227

Introduction

Fermenting your own food may seem like a daunting and unattainable process to take on at home. We have to admit, we were there one day too. But after you wrap your head around it, understand fermenting's history and methods, and know why fermented foods are so beneficial to your health, you may just reconsider. And because you're reading this book now, you're likely there already.

Coauthor Marni: As someone whose philosophy is rooted in natural nutrition and plant-based eating, each step of my journey with food has been a true evolution. I've taken on each layer with true determination. When I first made the choice to mostly eat plant-based food, it only became natural to take on all the things that came with eating a more wholesome, natural way. These include sprouting, juicing, blending, and — the most intimidating of the bunch — fermenting.

I resisted fermenting for years, thinking it was too difficult, but when I started with my first batch of sauerkraut, I had reached new levels. The concept of it was a true novelty. It was so simple and so rewarding to preserve your own food. Of course, experimenting with all the different plant-based foods was my true mission, so it was only natural to start taking some of my existing recipes and altering them to become more nutritionally active and alive.

Coauthor Amy: I can't remember a time when food wasn't an activity in my family. Sauerkraut, corned venison, pickled eggs — these were the foods of my people. It wasn't until I became a teenager that I realized no one else knew what in the world I was eating! From this upbringing, I developed a taste for fermented foods.

There's something intriguing about taking a single food and completely changing the taste and texture in a simple way. The smells and tingle of good fermented nourishment always remind me of being loved and cared for. I bring that same sense of self-reliance and health to this book. Each time I teach someone about the benefits of fermenting foods, I know that it opens up a whole new world of delicious taste that can come from their own backyard. There's no need to travel to exotic locations to experience delightfully tasty, nutritious food. You can create a wide variety of recipes with the simple ingredients you have on hand and a little time.

About This Book

This book breaks down the whole process of fermenting, from how to get started to what equipment and ingredients you need to discovering all the different foods that you can ferment. You'll be quite amazed at what takes place during the process of fermentation.

Here are a few of the topics that we explore:

- The basics of fermentation
- How to get started
- The benefits of fermented foods
- All the different types of foods that you can ferment: vegetables, fruits, grains, beans, meat, dairy, and beverages

On a technical note, you may notice that some web addresses in this book break across two lines of text. If you're reading this book in print and want to visit one of these websites, simply key in the web address exactly as it's written in the text, pretending as though the line break doesn't exist. If you're reading this as an e-book, you've got it easy — just click the web address to be taken directly to the website.

Foolish Assumptions

This book is for anyone interested in exploring the world of fermented foods. They're fascinating to learn about and healthy to boot. So if you're looking to take your health up a notch and take your foods to the next level, then this book is for you!

We expect that you have an interest in the topic and that you probably fall into one of these categories:

- You want to learn the basics of fermentation.
- You want to preserve your foods using natural methods that don't involve cans, vinegar, or other methods of food preservation.
- Your digestion is weak and you're looking for a natural solution through foods to heal your gut.

✔ You're interested in getting back to your roots and into the kitchen.

✔ You're looking to make your own homemade fermented beverages and foods so that you can avoid the not-so-healthy options in the supermarket.

Icons Used in This Book

This book uses *icons* — small graphics in the margins — to help you quickly recognize especially important information in the text. Here are the icons we use and what they mean.

This icon appears whenever an idea or item can save you time, money, or stress as you add more fermented foods to your diet. These include cooking and shopping tips and ideas for incorporating fermented foods into some of your existing meals.

Any time you see this icon, you know the information that follows is so important that it's worth reading more than once.

This icon flags information that highlights dangers to your health or well-being.

When the discussion veers off into the realm of too technical or downright nerdy, you'll see this icon.

Beyond the Book

In addition to the material in the print or e-book you're reading right now, this product also comes with some access-anywhere goodies on the web. Even though we try to give you as much info as possible in this book, you'll likely want to find out more. Check out the free Cheat Sheet at www.dummies.com/cheatsheet/fermenting.

In addition, you can read interesting companion articles that supplement the book's content at www.dummies.com/extras/fermenting. We've even included an extra top-ten list, which *For Dummies* readers seem to love.

Where to Go from Here

For Dummies books are organized in such a way that you can surf through any of the chapters and find useful information without having to start at Chapter 1. Naturally, we encourage you to read the whole book, but this structure makes it very easy to start with the topics that interest you the most.

If you're looking to gain an understanding of fermentation and its roots and what you need to get started, take a look at Chapters 1, 2, and 4. If you're trying to understand why fermented foods are so good for you, check out Chapter 3. If you're vegetarian, vegan, or generally plant-based in your diet, you may want to refer to Part III, which covers all things plant-based.

Chapters 11 through 13 cover animal-based fermenting. So if you're looking to ferment dairy, meat, or fish, these chapters are for you. The book also has several chapters on fermented beverages, from healing beverages in Chapter 14 to alcoholic beverages in Chapters 15, 16, and 17.

No matter where you go in *Fermenting For Dummies,* you're sure to discover a lot and gain a healthy attitude toward fermented foods!

Part I
Getting Started with Fermenting

In this part . . .

- ✔ Get familiar with the basics of fermentation.
- ✔ Discover traditional to modern practices of fermentation.
- ✔ Find out the difference between pickling and fermentation.
- ✔ Figure out the items you want to ferment.
- ✔ Understand why fermented foods are so good for you and your gut.

Chapter 1

In the Beginning: Fermenting Roots

*B*efore the days of refrigerators, people had to do something to keep their foods from going bad. Fermentation is one of those incredible preservation methods still used today. You can preserve foods in so many different ways: You can freeze them, can them, dry them for storage, or ferment them. These days, few people know and love the art of fermentation, but it's an art that has existed for many years past and, when you discover it, a world of splendor opens up!

Fermented foods are returning to the modern kitchen. The art of fermentation precedes history and happens by capturing and controlling the growth of bacteria, molds, and yeasts, and falling in love with the presence of lactobacilli found on the surface of all things. You'll discover more about the importance of these healthy living microbes in fermentation as you read on.

Fermentation is a unique, natural, and fun way to preserve your food, discover new flavors and recipes, and go on a mind-bending adventure through various cultures and through an ancient history of food that has existed for centuries around the world. If you're lucky, fermentation can even act as a tool for self-discovery and a vehicle for self-exploration in health and healing.

Getting Familiar with Fermentation

Fermented foods are all around you. You may not realize it, but you're likely already a consumer of one or more fermented food products. Have you had any sourdough bread, soy sauce, tofu, yogurt, cheese, or a glass of cider or wine lately? Does your sandwich come with a salty pickle or some sauerkraut on the side? You can thank the process of fermentation for these items.

Fermentation turns sugars to alcohol or other acids using yeast and bacteria. The chemical change often involves increasing the acidic environment and develops in places without oxygen (*anaerobic* conditions). It's a low-cost, highly efficient way of preserving foods.

Fermented foods have existed for centuries as populations around the globe learned how to capture the slow decomposition process of organic materials and preserve them by adding salts, sugars, or yeasts. They controlled mold and promoted good bacteria with the intention of maximizing the shelf life of their foods, enhancing flavors, or gaining health benefits.

Getting to know the art of fermentation also gets you familiar with the beauty of bacteria and its desirable presence in your food products. The changes caused by fermentation can be both good and bad. When fermentation occurs naturally, the food can smell or taste "off" (think of sour milk), but when you control the fermentation process, you can actually have some incredible results! When you execute fermentation processes properly, something that could have turned rotten instead turns into a consumable product. That's right — bacteria, yeasts, and molds will soon become your new best friends.

When fermenting foods, the key to developing the perfect environment and flavor and gaining all the great health benefits is to be confident, experiment, and do your best to create the utmost environment for fermentation, with proportional ingredients to support its growth. Some recipes are more challenging than others or require longer fermentation time, but plenty of fun and simple recipes are out there for beginners.

Fermentation throughout History

Food can give you insight into cultural and culinary traditions from around the world. Every part of the world has had a fermented food to be proud of. From beverages and breads to vegetables and fruits to meats and milk, there is often a whole culture and ritual behind these fascinating fermentations. This section pulls back a historical veil and reveals some rhymes and reasons behind this unique food process.

Mesoamerica: Calling all chocolate lovers!

Fermentation is essential to making delicious and tasty chocolate. The history of chocolate began with the Mayan civilization. The cacao tree grows in the tropics and produces a long fruit pod that, when ripe, is yellowish in color and contains anywhere from 20 to 30 cacao beans, or seeds, surrounded by a delicious white, fruity pulp. The seeds are left inside the white pulp to ferment and begin changing the chemical compound and releasing the flavor of chocolate that you know and love into the beans. These seeds are what is harvested and processed to make chocolate. Some cultures used the fruit pulp alone to make a fermented, slightly alcoholic drink consumed by Aztec warriors and aristocrats. Although some chocolate is made using unfermented cacao beans, the most flavorful and least bitter chocolate is born from fermentation. Cacao beans were so valuable in Mayan civilization that they were even used as a form of barter and currency! (See Chapter 14 for a drink recipe that uses cacao.)

Africa: Turning toxins into edible tubers

The *cassava root* is consumed in many parts of the world but has a strong presence in Africa. It is very rich in starch, a great calorie filler, and a relatively cheap market item. This staple food is abundant locally and cooked in many different ways. Deep-fried, steamed, boiled, or fermented, cassava can be sweet or savory. It needs to be fermented or cooked because it contains an amount of cyanide that's unpalatable and toxic to human consumption. *Gari* is the name for the common fermented cereal made from cassava, which could be compared to North American oatmeal, only fermented. (See Chapter 10 for notes on how to prepare cassava.)

Asia: Thirst-quenching and candied culture

Kombucha is one of the strangest looking fermentations, as it is done using a *SCOBY* (symbiotic colony of bacteria and yeasts) and appears rubbery in nature when growing. When placed in the correct environment, the combination of a SCOBY with tea and sugar creates an ancient health drink, *kombucha*, a fermented tea that is said to have originated in Central Asia. When drunk in moderation, kombucha has a wide range of health benefits. In some cases, the SCOBY alone is even candied by adding lots of sugar. Today, kombucha is becoming widely recognized among health food shops and within new-age environments. (See the nearby sidebar, as well as Chapter 14, for more on kombucha.)

Eastern Europe and Russia: Bubbly fruit kvass

Kvass is the Eastern European version of Asian kombucha. It's a fermented beverage that's most commonly made from rye, though other yeasts and fruits can be used. It has a low alcohol percentage and has been a common drink in Eastern Europe, and especially Russia, for centuries. In many cases throughout their patriotic history, people have chosen kvass over Coca-Cola! (See Chapter 14 for a kvass recipe.)

Japan: The sensational soybean

The soybean has become a widely cultivated and commonly distributed fermented food product. Tofu, tempeh, miso, and soy sauce are among the most recognized fermented soy goods, which originated in East Asia. The soybean itself has been cultivated around the globe and is a major industrialized food that serves populations worldwide. Although many people have problems with soy allergies, in moderation the fermented soybean (covered in Chapter 9) can actually enhance digestibility, reduce gas and bloating, and add beneficial flora to a person's diet.

Alleviating digestive discomfort

The exciting thing about being a culinary nutritionist is that I (Marni) get to inspire people and help them improve their health. Every client presents a different challenge, and we work together to find unique solutions to suit that individual's body and lifestyle needs.

Many of my clients come to me with digestive issues, and one client in particular had been suffering from digestive discomfort for years. She had a list of common symptoms: bloating, gas, and irritability. What was happening to her gut? I suggested she be daring and try something new. I suggested she try making her own kombucha or at least buy some to include in her diet every day.

The results were incredible! After sipping just a half cup every day with lunch for a week, her bloating subsided, her energy increased, and she felt significantly better. As you can see, just a small amount of fermented foods can have a profound effect on the body!

North Africa and Morocco: When life gives you lemons

Morocco tells a different story of the lemon. Lemons may be the last thing you thought of putting into your mouth whole, but with the magic of fermentation, you can eat them rind and all. When lemons are quartered, salted, and stuffed into jars left to ferment, they transform into a zesty treat. You can leave them in saltwater brine for years (see a recipe for preserved lemons in Chapter 6) and then use them in stews and sauces or to add a zesty kick to any recipe.

How Can Something Rotten Be Good for Me?

In her book *Nourishing Traditions,* Sally Fallon says that the proliferation of *lactobacilli* in fermented vegetables enhances their digestibility and increases vitamin levels. These beneficial organisms produce numerous helpful enzymes, as well as antibiotic and anti-carcinogenic substances. Their main byproduct, *lactic acid,* not only keeps vegetables and fruits in a state of perfect preservation but also promotes the growth of healthy flora throughout the intestine.

Fermented food helps turn those hard-to-digest substances into digestible ones and even comes chock-full of vitamins and minerals.

It may be deceiving that a food that has seemingly started to ferment can be good for you. Yes, the line can seem quite thin between rotting and fermenting, but as you get to know the art of fermentation, you'll quickly discover the difference. Food that is rotten has already become useless and inedible. It can smell bad, be moldy, and can certainly harm one's health. Fermented foods actually prevent rotting, can even be safer to eat than fresh food, and last much longer before they're considered truly spoiled. Fermenting foods can enhance the foods' flavors — you'll grow to love the new smells, strange fizzes, and interesting looks.

Fermented foods offer some amazing health benefits. They can

- Improve your digestibility
- Help you better absorb more vitamins and minerals

✔ Lower your risk of eating spoiled foods or getting food poisoning

✔ Reduce your risk of cancer and other diseases

For more information on how fermenting foods increases the nutrients in the food and the digestibility of your gut, see Chapter 3.

Fermenting Essentials

The fundamental things you need to ferment foods are often the same, but there are many variations of those ingredients that can change your results.

✔ **Fermenting containers:** Fermented foods must be made without the presence of oxygen or spoilage will occur. A good fermenting container is essential to your success. Use a sturdy container that's large enough to hold your fermented goods. Containers are best made from glass, like Mason jars, or nonreactive materials, such as a crockpot made from ceramic or a well-cleaned plastic bucket. The key to fermentation is creating an anaerobic, or oxygen-free, environment by sealing out any outside air.

Look for fermentation jars with an airlock seal that allows gas to escape but no air to get inside, though in some recipes a weighted lid will do the trick.

✔ **Lactobacilli:** These naturally occurring bacteria are essential to the fermentation process. These good bacteria have been proven to fight intestinal inflammation and help create a healthy gut. They also enhance the flavors and digestibility of fermented foods — they're the invisible workers that make your food ferment!

✔ **Salt:** Salt can kill any bacteria that may cause illness. It does this by creating a less inhabitable environment by removing water from the plant cells. Salt also helps enhance the flavors of food. It can reduce sweetness or bitterness in foods, a desirable thing for your recipes!

✔ **Spices and herbs:** You add herbs and spices to your fermented foods to create unique recipes. Think of adding ginger to your kombucha, cranberries to your sauerkraut, or caraway seeds to your pickled goods!

✔ **A starter or a culture:** Many fermented recipes ask for a starter or a culture. No, we're not looking for you to adopt a new way of life; this type of culture is one full of existing microbial life. A fermentation starter can come in the form of a dried powder, yeast, or a wet substance and is essentially used to boost the food's flavor and the digestion process.

You can get good results using a kick-start from a previous batch to accelerate the fermentation process. You can purchase starters or, depending on the product, reuse them from other food products like sourdough or yogurt.

✔ **Sugar:** You use sugar to help preserve foods when salt would be undesirable. (Imagine making jam or kombucha with salt. Yuck!) Most commonly used in wet brine, sugar can include cane sugar, honey, or maple syrup. (Refer to Chapter 6 for info on sugar substitutes.)

✔ **Time:** Every good fermented food product needs time. The tiny microbes will work to turn those starches into sugars and alcohol and will only slow down if you place them in cooler temperatures. Depending on the end product, you'll leave your ferments anywhere from two to seven days, or longer! Check your recipe and taste your food according to the flavors you desire.

Pickling (and How It Differs from Fermenting)

There are so many different kinds of pickles! Pickles are generally associated with the traditional cucumber in brine, but you can pickle all kinds of things, from fruit and vegetables to meat, fish, and eggs. In India, some of the most popular pickles are made from mango and lime. In Europe, you'll find pickled herring, olives, and beets. From Asia to Europe, the world of pickling is vast and varied.

Pickling is the process of preserving food using a brine (saltwater) solution. The salt in pickling acts on the food by drawing out the water from its cells and kills any bad bacteria that may spoil the food. Pickles are often added to a meal to help aid with digestion, giving your body that extra bit of *Lactobacillus acidophilus* it needs to restore some healthy gut and intestinal flora.

So, what's the difference between pickling and fermenting? Fermenting and pickling can seem very similar, but they're not the same thing. The process that occurs inside the brine is called fermentation, but the act of making brine and placing food into the saltwater solution is called pickling. Pickling also usually requires added heat through a canning process, whereas fermented goods can sit out on your shelf and don't require heat. Fermented foods thrive in anaerobic conditions and make use of naturally occurring "good" bacteria submerged under the saltwater. Fermented foods have a bit of a tangy flavor, while pickled goods taste salty or vinegary all the way through.

The role of salt in fermentation is to help draw out water from foods and make a salty living environment so bad bacteria have little chance of survival. A brine is created in fermented recipes to preserve fruits and vegetables or other food products. Making brine can be a bit of an experiment, and the salt measurements sometimes depend on your personal preference. Remember that a little salt goes a long way! Some brine even contains a bit of sugar to balance out the salty flavor.

Here are some general tips on pickling:

- ✔ Use the firmest, freshest vegetables or fruit possible.
- ✔ Scrub your veggies well before using them; any dirt or bacteria can affect them.
- ✔ Make sure you clean and sterilize all your pickling supplies.
- ✔ Choose unrefined salt for the highest-quality and best-tasting pickles.
- ✔ Keep your vegetables submerged under the brine (salt solution).
- ✔ Wait the recommended period of time before eating your pickled goods.

A Quick and Easy Intro to Fermenting: Making Sauerkraut

Before we end this chapter, we want to show you just how easy it is to experience the wonders of fermented food. One of the most popular forms of fermenting and pickling is done with the common cabbage. For centuries, cabbage has been made into kimchi in Korea and sauerkraut in Germany. There are many different types of sauerkraut. Sauerkraut is an incredible recipe that uses the process of wild fermentation, meaning that no starters are needed. The natural bacteria living on the plant life are responsible alone for the ferment. It doesn't get any easier than that, which makes sauerkraut a perfect recipe for fermentation beginners!

Here are the items you need:

- ✔ One head of cabbage (any color your heart desires)
- ✔ Sea salt
- ✔ A well-sharpened knife and a cutting board
- ✔ Cleaned and sterilized fermenting containers; a glass bowl, wide-mouth Mason jars, ceramic bowls, or plastic buckets will all work just fine

Sometimes people use a weighted object to cover sauerkraut as it ferments. Whether you use one depends on the fermenting container you're using. For sauerkraut, large ceramic bowls or wide-mouth jars are certainly handy. Keeping your cabbage submerged under the wet brine is most important. A weight, something as simple as a heavy plate or a plate with a stone on top, helps to put pressure on the top of your container and pack down your cabbage.

Try to ensure that the plate fits snugly around the circumference of any vessel you use. You want to allow the gas to escape but minimal oxygen to get inside that may cause mold to build on your sauerkraut. If you notice mold, you can scrape it off and still eat the cabbage below. The fermentation process and all the salt will kill off any unwanted microbes.

After you gather your materials and equipment, here's how you make sauerkraut:

1. **Chop the cabbage into fine, thin slices.**

2. **Place the cabbage into a large bowl or plastic bucket and add an even amount of salt to cover the cabbage.**

   Two to four tablespoons should suffice, but feel free to add according to your taste and preference.

3. **Mix the salty cabbage and squeeze the vegetable until liquid begins to emerge from it.**

   The act of salting the cabbage draws out enough water. You'll get to see this over time.

4. **Place the squeezed-cabbage tightly into a Mason jar or other fermenting container.**

   Press it forcefully down into the bottom of the container so it's packed well. You don't want to have any air bubbles. Ensure that the brine (salt-water) is covering the cabbage.

5. **Leave the cabbage submerged under the brine.**

   If you want to add a small amount of water you can, but this isn't necessary.

6. **Seal the jar or cover your fermenting container with a heavy weighted object.**

   A comfortable fermentation environment is at or just below room temperature. Some people suggest covering the top of your sauerkraut with several full cabbage leaves before placing the lid on top. Try this for yourself and see what works best. Leave your jar sitting in a room slightly colder than room temperature for three to five days.

7. **Check on your fermented goods daily.**

   Pressure will begin building on the inside of your container, and you may see some water overflowing. You can taste the changes over time and adjust the salt or water content as needed (add some water if you find the brine isn't covering the cabbage). If the cabbage you use to ferment isn't fresh, the water content is naturally going to be less.

8. **After several days, your sauerkraut should be ready to eat.**

   When you've achieved the softness and salty flavor you desire, serve it as a side dish or just eat it alone. It will last for months in your refrigerator.

Chapter 2

The 4-1-1 on Fermenting

*I*n this chapter we explore why people ferment foods. There has to be a good reason — it has been done for years! We discuss everything from the process itself to which organisms are involved to what you need to make successful ferments at home. Just knowing that you can make your food last longer and that fermented foods are good for you should be enough to get you excited!

We're confident that this chapter will familiarize you with fermenting basics so that you can get started right away.

Why Ferment?

You ferment food to change the food in a controlled way, into something completely different. Fermenting is used in common foods and beverages like beer, yogurt, vegetables, and even meats. After a food has been fermented, it not only tastes different but also *is* different on a molecular level.

Fermenting releases some nutrients that would otherwise be unavailable for digestion and creates an environment for *probiotic,* or beneficial, bacteria to thrive. These bacteria already live in a healthy digestive tract and help break down food properly. Antibiotics and digestive illnesses can reduce the number of good bacteria, making you susceptible to illness. Your immune system includes your digestive tract, and improving digestion can only help overall health.

When making fermented foods, you actually sour the food rather than pickle it. Because fermenting, or souring, is a controlled decay of the food, you must follow all food safety rules. Always ferment in the most sanitary conditions possible and use very fresh, exceptional food. Adding fermented food to your pantry is not only tasty but also healthy.

Fermenting is simply decaying with style. There's a fine line between going bad and fermenting, so the rules of a good fermenting system are stringent. These rules become easier to adhere to as you become proficient, and the benefits are worth it.

Getting Acquainted with the Good and Bad Guys

Mold, yeast, bacteria, and enzymes are the four spoilers. *Microorganisms* (mold, yeast, and bacteria) are independent organisms of microscopic size. *Enzymes* are proteins that exist in plants and animals. When any one or more of the spoilers have a suitable environment, they grow rapidly and divide or reproduce every 10 to 30 minutes! With this high-speed development, food can spoil quickly. Some types of spoilage can't be seen with the naked eye (like botulism), while others (like mold) make their presence known visually.

Food spoilage is the unwanted deterioration in canned or preserved food that makes it unsafe for eating. Ingesting spoiled food causes a wide range of ailments, depending on the type of spoilage and the amount of food consumed. Symptoms vary from mild, flulike aches and pains to more serious illnesses or even death. This is why we stress following the directions exactly in any proven recipe for fermenting, and knowing the signs of problems.

Living microorganisms are all around — in your home, in the soil, and even in the air you breathe. Sometimes microorganisms are added to food to achieve a fermented product, like beer or bread (for leavening). They're also important for making antibiotics. The point? Not all microorganisms are bad — just the ones that cause disease and food spoilage.

Mold

Mold is a fungus with dry spores. Poorly sealed jars of high-acid or pickled foods are perfect locations for these spores to set up housekeeping. After the spores float through the air and settle on one of their favorite foods, they start growing. At first, you see what looks like silken threads, then streaks of color, and finally fuzz, which covers the food. Processing high-acid and pickled food in a water-bath canner destroys mold spores.

Don't eat food that's had fuzz scraped off of it. This was thought safe at one time, but not anymore. Mold contains carcinogens that filter into the remaining food. Although the food appears to be noninfected, ingesting this food can cause illness.

Yeast

Yeast spores grow on food like mold spores. They're particularly fond of high-acid food that contains lots of sugar, like jam or jelly. They grow as a dry film on the surface of your food. Prevent yeast spores from fermenting in your food by destroying them in a water-bath canner.

Bacteria

Bacteria are a large group of single-celled microorganisms. Common bacteria are staphylococcus and salmonella. Botulism, the strain of bacteria to be most concerned with in canning, is the most dangerous form of bacteria and can be deadly. It's almost undetectable because it's odorless and colorless. Botulism spores are stubborn and difficult to destroy.

Because botulism is undetectable without laboratory testing, handling and preparing canned food according to the most updated guidelines available is essential.

Botulism spores hate high-acid and pickled foods, but they love low-acid foods. When you provide these spores with an airless environment containing low-acid food, like a jar of green beans, the spores produce a toxin in the food that can kill anyone who eats it. The only way to destroy these spores in low-acid food is by pressure canning.

Enzymes

Enzymes are proteins that occur naturally in plants and animals. They encourage growth and ripening in food, which affects the flavor, color, texture, and nutritional value. Enzymes are more active in temperatures of 85 to 120 degrees than they are at colder temperatures. Enzymes aren't harmful on their own, but they can make your food overripe and unattractive while opening the door for other microorganisms or bacteria.

An example of enzymes in action occurs when you cut or peel an apple. After a few minutes, the apple starts to brown. We call this *oxidization.* Eventually, oxidized food will spoil.

The Mechanics of Fermenting

Fermenting is a glamorous term for decaying your food in a controlled method. Yes, you're allowing your food to decay. Your food is decaying from the moment you pick it. If you ferment correctly, you create the perfect environment of temperature, liquid, and oxygen (or lack thereof) to change your food into a delectable, sometimes healthier food.

If you've never fermented, don't worry. The piquant taste is something that's enjoyable for most people and unforgettable to anyone who tries it. Fermenting is a technique that you can try without a lot of financial investment in equipment and specialty supplies. It has always been a simple way of preserving foods, simple enough to do in the farming kitchen.

If you're eating locally, your choices for variety are limited. Fermenting is an easy way to expand the flavors and textures of the same short list of foods and keep your diet interesting. Fermented foods also provide nutrients that aren't available in fresh or nonfermented foods. Consider the fermented foods to be an extension of your food pantry.

Fermenting can't increase a food's quality. To develop a perfectly fermented food, you must start with perfect produce, fresh meats, and milk. Any cuts or bruising is where decay is happening, and you don't want to give unwanted bacteria any chance to be present in the fermenting container. In the case of meats and milks, ferment them as soon as possible, as they also degrade over time.

Salt curing and drying versus fermenting

There are other methods of food preservation besides fermenting. *Salt curing* and *drying* are commonly heard terms that may be confusing to sort out from the fermenting process.

✔ **Salt curing:** This method of preservation is easily recognized in the cured favorite: bacon. Dried beef and real country ham also have the strong taste from curing's main ingredient: salt.

You can add other ingredients to the salt mixture to help cut the strong salty taste and add additional flavors, but you can't entirely hide that salty flavor. Plain salt curing has fallen out of favor over the years. It successfully cures the meat but imparts a harsh flavor. Some meats are still preserved country-style, or plain salt cured, but they must be soaked before cooking to remove the excess salt and make the meat palatable.

✔ **Drying:** This is one of the most delicious techniques for meat storage. Drying a piece of meat properly only enhances the flavor and creates a uniquely flavored product that truly satisfies. Drying is how jerky is made. The flavoring you add becomes concentrated as the meat dries. This process results in a chewy, delicious snack that packs a protein punch. From sweet to spicy, you'll find a never-ending array of flavor profiles for jerky. You can dry meat in an electric dehydrator, an oven, or by sitting it out in the sun.

An electric dehydrator is the best and most efficient unit for drying, or dehydrating, food. Today's units include a thermostat and fan to help regulate temperatures. You can also dry food in your oven or by using the heat of the sun, but the process takes longer and produces inferior results to food dried in a dehydrator.

Vinegar and acids versus fermenting

True vinegar is more than the flavor behind a dill pickle. Vinegar in its raw form is actually a living food, containing the beneficial bacteria you need to digest your food properly.

Several types of vinegars are used in cooking. Some of them are commonly found in a well-stocked pantry, while others are usually found in a specialty store or the specific country where they're used.

For the most part, the vinegar found on your grocery store shelf isn't a true vinegar, or it has been pasteurized to kill the beneficial bacteria inside. In fact, many vinegars are simply flavored and colored white vinegar. Be sure to read the label and look for the words *raw* and *unpasteurized.*

Adding vinegar to a food is *pickling,* not fermenting. You can enjoy pickled foods and fermented foods, but they're not interchangeable.

Here are some of the vinegars you'll find commercially:

✔ **Apple cider vinegar:** Brown in color, this vinegar is available raw, with the mother still visible. It's usually a naturally fermented product, but read the label to see whether it's simply white vinegar that has been colored to look like cider vinegar.

If you buy a bottle of pasteurized apple cider vinegar, you can add the mother to it and replenish the beneficial bacteria that were killed during the pasteurization process. Simply add a tablespoon of raw apple cider vinegar to the new bottle of pasteurized vinegar and allow it to sit for about a week. You'll see the mother beginning to form on the bottom of the new vinegar bottle.

- **Balsamic vinegar:** Made from grapes, this almost black vinegar is aged for many years. The flavor is complex, and it's used in dishes that can showcase the unique flavor.

- **Fruit vinegars:** These vinegars are made from the fruit that's allowed to ferment. They have a delicious fruity flavor and are wonderful for cooking.

- **Rice vinegar:** This vinegar is made from rice, the color denoting the type of rice used in the production. Rice vinegar is mild tasting and never overpowering.

- **White vinegar:** Actually a distilled product, this vinegar contains 5 to 8 percent acetic acid in water. This type of vinegar is used for many kitchen preparations, like pickling and baking, because of its plain flavor. White vinegar has no beneficial bacteria.

Canning versus fermenting

Don't mistake fermenting food for canning food. *Canning* is the process by which food and liquid is placed into heat-safe jars and then heat is applied to destroy any microorganisms that can cause food spoilage. All foods contain these microorganisms. During canning, the air is driven from the jar and a vacuum is formed as the jar cools and seals. This prevents microorganisms from entering and re-contaminating the food.

Fermented food is the process by which food is submerged under a liquid, and in this anaerobic environment, beneficial bacteria are allowed to grow. These bacteria would be destroyed by the high temperature of canning.

Refrigerating fermented food

Placing your fermenting food in the refrigerator slows the fermenting process down considerably. Some foods, like meats and eggs, are fermented for a short time at room temperature to allow the bacteria to begin the process, and then the fermenting is moved to the refrigerator for days and even weeks, to continue to work at a slow rate. This gives the beneficial bacteria time to work while prohibiting dangerous or unwanted bacteria from starting to grow.

What happens if you don't ferment your food?

If you're still on the fence about fermenting food, consider this: All food decays. Simply relying on food being the freshest and most wholesome it can be usually results in overbuying or wasting food because you can't eat it before it spoils.

If you're trying to stick to eating local, seasonal foods, by midwinter, you have little variety to your diet, and the nutritional benefit of fresh foods decreases as they're stored and degrade over time.

By fermenting, you preserve food in a way that increases its nutritional content and changes the flavor enough that you can use it in a whole new way.

Placing fermented food into the refrigerator is also a good trick for keeping a fermented food from becoming sourer than wanted. When your fermented food tastes just the way you like it, place it in the refrigerator to keep it tasting the same.

Freezing fermented food

You can freeze fermented food for storage, but doing so isn't optimal. When fermenting, if you place food in the freezer, it won't continue to ferment. Fermented food that's frozen also changes texture after it's thawed. Overall, go ahead and freeze your fermented food if you find that you have excess, but it should be a last resort.

If you find that you often have to freeze your excess fermented food, you should decrease the amount of food you make at a time. Also consider fermenting at a cooler temperature so the ferment time takes longer, possibly giving you enough time to finish the first batch.

Fermenting foods and drinks

Much more than beer-making, fermenting is the technique behind the sour tang of sauerkraut, vinegar, and yogurts. Fermenting is also the perfect beginner's preserving technique because it takes very little time and requires a short list of ingredients.

Use recipes from reliable sources or ones that you've already made successfully. Follow your recipe to the letter. Don't substitute ingredients, adjust quantities, or make up your own food combinations. Improvisation and safe food preservation aren't compatible. This approach also means you can't double your recipe. If you require more than what the recipe yields, make another batch.

Preventing food spoilage is the key to safe fermenting. Over the years, people who ferment their foods have perfected the most modern methods and recipes for the best outcome. When you follow up-to-date guidelines exactly, you'll experience little concern about the quality and safety of your fermented foods.

Avoiding spoilage

In addition to choosing the right preservation method, follow these steps to guard against food spoilage:

- **Don't experiment or take shortcuts.** Use only tested, approved methods.

- **Never use an outdated recipe.** Look for a newer version. Don't update the directions yourself. Check the publishing date at the beginning of the recipe book. If it's more than five years old, find a newer version.

- **Use the best ingredients and follow recipes to the letter.** Because botulism and other food-borne bacteria can be tasteless and odorless, this advice is your best defense.

- **When curing meats, always use the proper amount of cure, which includes nitrates/nitrites (pink salt).** These ingredients have gotten a bad rap in the media, but the truth is, without them, your preserved meat can be silently dangerous.

Chapter 3

The Benefits of Fermenting

The process of fermenting foods has a long history and is proving to be one of the most healthful ways to consume foods. Fermented foods offer an array of health benefits for digestion, immunity, and overall well-being, while adding some dynamic flair, crunch, texture, and carbonation to your meals. This chapter explores some of the benefits of fermenting your food. We promise that you'll start to notice a difference in how you feel almost immediately!

The Lowdown on Lacto-Fermentation and How It Helps Your Body

Lacto-fermentation is a natural biological process in which sugars and starches are converted in lactic acid. This process happens without the presence of oxygen, which in itself is what makes lacto-fermentation so unique. The organisms that come to life as a result of lacto-fermentation are why it's one of the most beneficial fermentation processes, because they help with so many of the body's functions, from cell formation to circulation to balancing digestive acids. One of the main benefits of lacto-fermentation is enhanced digestion; the presence of lactic acid makes it easy for foods to be absorbed in the digestive tract.

Lactic acid helps balance the acidity in your stomach and directly aids your body's glands in secreting digestive juices to break down foods.

Boosting your health with vitamins and minerals

No heat is involved in the process of fermentation, so the vitamins and minerals in the food are left intact. The best part is that beyond just preserving the vitamins and minerals, fermentation can actually create new ones. It also helps make the minerals in the food easier to absorb, pre-digests the food, and neutralizes harmful food components, making it easier for your body to absorb nutrients all around.

Each food is unique, which means that the fermentation process interacts with it uniquely and allows more specific nutrients to be available, depending on the food item. However, vitamins C and B are increased exponentially during the process of fermenting most foods.

Vitamins and minerals are the most essential nutrients that your body needs to thrive. You need them for almost every function in the body, from cell and tissue growth to repair to metabolism.

- ✔ Vitamin C is crucial for tissue growth and connectivity and is extremely important during pregnancy and to boost the immune system.

- ✔ Your body uses B vitamins to make energy from food and to make new red blood cells. These vitamins aren't stored in your body, so you must replenish them on a regular basis, preferably from whole-food, plant-based sources.

- ✔ Minerals are also essential to good health, including calcium, magnesium, iron, and potassium, among many others. They're readily found in various sources of green leafy vegetables, nuts, seeds, and beans.

Loading up on enzymes

Enzymes are substances, usually proteins, that act as catalysts to bring about specific action in the body, such as the digestion of particular nutrients.

Your body has thousands of enzymes, but it runs on only two types of them:

- ✔ *Digestive enzymes* act as little workers that help break down the food you eat, especially the macronutrients, protein, carbohydrates, and fat.

- ✔ *Metabolic enzymes* are made within the body and can only function within the body for all the reactions that take place that you don't even think about, such as aiding the functioning of glands, removing toxins, producing energy, and purifying the blood.

As you age, your body gradually produces fewer and fewer enzymes, making it more difficult for your body to digest food. And the fewer digestive enzymes your body makes, the fewer metabolic enzymes it creates to keep your body functioning optimally.

So the goal is to gain as many enzymes from food sources as you can. This is the optimal way to maintain health. Foods that are left uncooked, and not only raw but fermented, promote this state of well-being and optimal cellular and metabolic functioning. Thus, people need to eat more fermented foods, and eat them more frequently.

Here are three ways you can get enzymes from fermented foods:

- ✔ They're already present in the food, in raw foods like fruits and veggies.
- ✔ They're produced by the microorganisms doing the fermentation.
- ✔ They're produced by microorganisms that are around but aren't involved in the primary fermentation.

All that being said, fermented foods usually contain more enzymes after being fermented than they did before. This can mean a world of difference to someone who has problems digesting foods.

Adding only a tablespoon of sauerkraut to your meals can enhance the digestive process!

Aiding pre-digestion

When a food is pre-digested, the difficult work that your body would have to do to break it down is done for you. The process of fermentation allows beneficial organisms to feast on sugars and starches. What this ultimately means is that the carbohydrates, protein, and fats in the foods are broken down before your body can get to them.

So fermentation basically pre-eats your food — whether it's a vegetable, fruit, grain, or nut — and allows your body to use the nutrients in it much quicker and easier.

For example, in foods such as grains, the gluten is pre-digested. This is very good news for people who are sensitive to gluten because it can make the grains easier to consume. After a grain is soaked, sprouted, or fermented, it becomes a nutritional powerhouse and basically digests itself.

Celiac disease

Celiac (*see*-lee-ak) *disease* is a digestive condition triggered by consumption of the protein *gluten,* which is primarily found in bread, pasta, and many other foods containing wheat, barley, or rye. People with celiac disease who eat foods containing gluten experience an immune reaction in their small intestines, causing damage to the inner surface of the small intestine and an inability to absorb certain nutrients.

Note that the pre-digestion of grains doesn't help people who have celiac disease because they're completely intolerant of gluten and must avoid it (see the nearby sidebar for more info on celiac disease).

In the fermentation of beans, the resulting pre-digestion can really help the process of digestion because most people have a difficult time digesting the complex carbohydrates in them. So a fermented bean means a happy bean, and a happy bean means no gas for you!

Activating your foods

An incredible process takes place when foods are fermented. Just an overnight soak can sometimes do the job, but fermentation is what really completes this process. Foods such as grains, nuts, seeds, and beans contain a substance called *phytic acid.* If this isn't deactivated, then it can actually prevent mineral absorption in the digestive tract. Ultimately, this can lead to mineral deficiencies.

However, by fermenting your food, you activate the nutrients, making the food item more bioavailable in your body. This allows you to get as much nutritional value out of the food as possible.

Promoting probiotics

You may have heard of probiotics, and if not, you should get to know them. They're quite likely the best byproduct of fermented foods. Your body and gut mainly rely on them for efficient digestion, immunity, and overall health.

Probiotics are microorganisms consumed by the body for their beneficial qualities. *Gut flora* refers to the colony of microscopic organisms living in the intestines.

The gut, being the central spot where your food passes through at some stage of the digestive process, is where beneficial organisms like to thrive. In a healthy body, these organisms help absorb the nutrients from foods and help the body defend against and eliminate unwanted diseases, infections, and toxins.

If you lack beneficial bacteria in the gut, your body is left in a compromised state, opening up the possibility of a weakened immune system and weakened digestion, which can result in some uncomfortable situations.

Gut flora and beneficial bacteria naturally decrease as people age. However, other factors such as taking antibiotics, feeding on infant formula, eating a highly processed or acidic diet, and living near or being exposed to a toxic environment can all contribute to low levels of these beneficial bacteria.

That's why the introduction of basic fermented foods into your diet can provide such a drastic change to your health. Fermenting foods isn't just a hobby to preserve your food or expand your talents in the kitchen. It actually lends some extremely positive — even life-changing — results to your health.

Strengthening your immunity

Fermented foods boost immunity and protect the body from infection. Different types of bacteria have helped the body do this throughout history. Now it's about harnessing what can be done with food through fermentation and using that as a sort of prescription to enhance the immune system naturally.

If the immune system is compromised in any way, then the beneficial bacteria that are present get stripped away. This leaves a surface that's prone to infection, viruses, fungi, and toxins that weaken your immunity.

All in all, eating more fermented foods helps nourish the body with healthy beneficial organisms. When your gut is working efficiently, so is the rest of your body; they go hand in hand. Only with a healthy gut can you have a strong immune system. So drink that kombucha, eat that coconut yogurt, and crunch on that kraut!

How Fermentation Can Make You a Better Cook

If fermented foods didn't taste good, they wouldn't have remained as popular or as widely used as they have for centuries and centuries. We admit that some fermented foods may take time to get used to, like your first swig of kombucha. But by your second or third, you'll want that bubbly, natural fizz on a regular basis!

Overall, fermented foods can provide your meals with additional texture and crunch and loads of extra and often unique flavor.

As you see throughout this book, you can ferment a variety of foods, which means you can get an exceptional range of health benefits from pretty much all the food groups. These are just an added bonus to your daily diet.

Keeping it simple and easy

The fact that you can be your own food processor, enhancer, flavorer, and preserver of your food without a factory, chemicals, preservatives, or other unnatural things makes fermenting easy and extra special.

All you need is some basic equipment; you can be as low maintenance as just a few items or get yourself stocked with a few extra things. In Chapter 4 you can see the range of equipment that's required for different methods of fermentation. But in the beginning, make it easy on yourself and keep it simple!

Fermenting is something that even took me (Marni) a while to come around to. It seemed too intimidating, time-consuming, and difficult until I made my first batch of sauerkraut or fermented my first batch of water kefir. After you do it once, it becomes easy, and then you slowly get hooked! All I can do is encourage you to get started; now that you know how beneficial fermented foods are for you, what have you got to lose?

Following the seasons

Different times of year call for different methods of food preservation and, even more important, call for different foods. Many fruits and veggies either aren't available or don't taste very good at certain times of year, which is your indication to let them rest and wait until they're in their prime. If you stick with the seasons, you're doing yourself a service, and the fermentation process will also work to your advantage. Fruits and vegetables especially are at their best when their nutrients are just bursting, and this comes from being in season, organic, and harvested locally.

Figuring Out How Much and How Often

There are many beliefs and recommendations about how many fermented foods you should eat. Some people say to eat them many times a day, and others a few times a week. We say, just get them in!

Of course, if you're a novice, you may not even be making fermented foods yourself but rather buying a local kombucha or sauerkraut or enjoying the odd bowl of miso soup — and for you, perhaps this is enough. But making your own fermented foods allows you to have control over your food from beginning to end. Also, the closer the fermented food is to home, the more powerful and beneficial it is to your body.

As you start to expand and try more fermented foods, you actually begin to crave them (in a good way) and want to include them in most meals throughout the day. So here are some general guidelines to enjoy them:

- ✔ You get the most benefit if you include fermented foods in most of your meals throughout the week, if not every day and sometimes even every meal.

- ✔ When it comes to beverages like kefir, kvass, and kombucha, a few sips to a half cup is enough.

- ✔ A few tablespoons of fermented fruits or vegetables, like sauerkraut, are easy to get in daily.

- ✔ Condiments, dips, spreads, breads, grains, and beans may make up most of your meals during the day. You'll have to gauge for yourself how much to consume, but just know that although a little bit goes a long way, the more the better. So do whatever it takes to make fermented foods part of your meal planning.

Understanding the effects of ferments

If fermented foods are completely new to your system, you may experience mild to moderate symptoms, which are completely natural and healthy for your body. These symptoms can include gas, breakouts, headaches, and more. Just note that you have to give your body time to adjust and adapt. Nothing wrong is happening; in fact, only good things are happening. Your body is getting rid of toxins that need to be released. This is good, because the fermented foods are replacing those toxins with beneficial, life-enhancing organisms that will ultimately make you feel and look great!

Chapter 4

Getting It All Together

*I*n this chapter we explain what tools and equipment you need to make your own home-fermented foods. Luckily, you have a variety of equipment to choose from, and you can select what's right for you. This chapter also explains what you need to keep your tools clean. If your equipment isn't clean, your food won't ferment properly. We also cover some of the basic starting ingredients so that you can begin fermenting right away and without a large investment.

Assembling Your Equipment and Tools

You can use a variety of tools and equipment for fermentation. You can purchase your tools and equipment from specialty shops and sometimes at garage sales and secondhand stores. However, many people can make do with what they have lying around their kitchen and home, which is just fine too!

You can begin with basic, low-cost tools and equipment and expand into more expensive equipment as you get familiar with the fermenting process. The trick to successful fermentation isn't how much money you spend on tools and equipment but how much time you invest and how closely you pay attention to details.

Basic containers

You need several types of containers for fermenting. In some recipes, containers may be called *vessels.* Containers are made of various substances and come in many sizes. Different types of containers are better suited for different fermented foods. If possible, try to get lids with your containers. You won't always need the lid for sealing the jar, but when you do, it's next to impossible to improvise a tight-fitting lid.

Containers for fermenting can be as large as 20 gallons or as small as a quart. When you begin fermenting you want to stay with smaller containers, such as quart- and gallon-sized containers. The containers you use for fermenting sit in the same spot for a few days to several weeks, and if that's on a kitchen counter, it could make for some crowded conditions. Even when fermenting for a large family, using smaller containers is still prudent because doing so ensures faster consumption of one jar while the next one finishes working. Plus, smaller containers are simply easier to manage.

In my family of eight, the largest size container I (Amy) use is a gallon jar. I like to ferment in weekly-sized batches and have one ready to eat, one almost finished, and a third one just starting. The most common size I use is a quart, for our most-loved recipes.

Here's a look at some types of containers.

Metal containers

Commonly used for brewing, metal vessels provide a clean surface for fermenting any foods. Stainless steel is the best choice for a metal container. Enamel over steel is a good choice if the enamel has no chips on the interior surfaces. Chipped areas may rust, which isn't good for your food. Be careful to avoid metals like cast-iron, copper, aluminum, and tin, all of which can react with the acids in fermented food and give it a strange flavor or cause a color change. These metals can also leach into the food.

Don't use containers coated with Teflon or other nonstick coatings. The coating starts to peel away almost immediately, and then you consume it in the food. And these containers often have metal under the coating that's reactive to the acidic environment of fermentation, resulting in off flavors or colors in your recipe.

Plastic containers

You use plastic containers for some parts of the brewing process and for fermenting fruits and vegetables. Make sure you choose food-grade plastic that's BPA-free.

BPA products contain bisphenol A, a chemical compound that has been linked to health conditions such as infertility, cardiovascular problems, and diabetes.

Wooden containers

You typically use wooden containers for fermenting wine and beer. Wooden vessels are usually barrels, which come in varying sizes. Wood is difficult to sanitize, and we don't recommend it for most fermenting projects.

Glass containers

Glass jars are probably the easiest and most cost-effective vessels for fermenting. They don't hold any odor from previous contents, and you can sterilize them again and again. Canning jars are inexpensive and easy to find in most big-box stores. You can also repurpose jars from purchased products or use jars that you may have lying around the kitchen. The bonus with glass jars is that they come with lids. Although the lids aren't always high quality and may be tough to clean properly, the glass jar itself is worth saving. If I (Amy) had to choose one type of fermenting container, it would be glass.

Ceramic containers (crocks)

Using ceramic (also called *stoneware*) vessels is also a good choice. However, be aware that crocks can be heavy, something to consider if you have to move a crock that's full of product.

When buying ceramic containers from secondhand or antique stores for fermenting purposes, be wary of those that aren't marked food-safe. Many older crocks have glazes that contain lead, which can leach into your recipe and is dangerous. Also, be sure to check crocks carefully for cracks, which make the crock impossible to clean well and may cause liquid to seep out during the fermenting process.

Many hardware stores carry lead-testing kits that detect the presence of lead in an object. When in doubt, use caution.

Essential tools and utensils

Along with containers to hold the foods you ferment, you need some other tools and utensils to begin fermenting. Some of these are common kitchen items that you may already have. Other items can be repurposed or improvised if you don't want to purchase them. Here we explain what tools and utensils you need so you can gather them before you begin fermenting.

Weights

The purpose of a weight is to push the fermenting food down under the liquid used in a recipe and keep air out. Because the process of fermenting requires *anaerobic bacteria* (bacteria that work without oxygen), weights are important.

You can purchase weights from specialty stores or use a number of common household items, such as jars or plastic jugs filled with water and set on a flat glass or ceramic plate over the fermenting food. You can also use a clean brick or large rock to hold a flat plate down. Some people simply fill large zip-lock plastic bags with water to weigh down food.

Whatever you use as a weight, you must be able to clean it thoroughly and sanitize it, and it must be nontoxic because it comes into contact with the food.

Food thermometer

If you don't have a cooking thermometer, you may want to purchase a food thermometer. Knowing the temperature of fermenting food is important because the correct temperature is essential to proper fermentation and food safety. The temperature is also a handy indicator that you're successfully fermenting.

Choose a thermometer with a clip that goes over the side of the container so that the tip suspends in the liquid and doesn't touch the bottom of the container. A clip keeps you from having to fish through the container with your hands should the thermometer sink in the food.

Colander and strainers

Colanders and strainers are convenient for washing vegetables or straining off excess fluid from fermented milk products. Stainless-steel or plastic colanders and strainers are preferable to aluminum, which may react with some fermented food and cause a strange flavor.

Knives and peelers

You probably have knives and vegetable peelers in your kitchen. You use these to prepare foods for the fermenting process. You want a large chef's knife as well as paring knives for preparing foods.

A saying in the kitchen world says that a sharp knife is safer than a dull one. This is certainly true. Keep your knives sharp at all times to make clean cuts and to stay safe.

Cutting board

Every kitchen needs a cutting board, and you probably have one. If you don't, choose a metal or glass cutting board because they don't hold bacteria and they're easy to clean. If you use a wood cutting board, make sure to scrub it well before using. Clean all cutting boards with hot water and soap between cutting raw meat and cutting produce to avoid cross-contamination of food with harmful bacteria.

Mark your cutting board in an inconspicuous place, dedicating one side for produce only and one for meat only.

Measuring utensils

You need a good set of measuring spoons and cups for measuring the food, spices, and salts you ferment. These should be easy to sanitize.

Stirring and scooping tools

You need some spoons for stirring your fermenting foods. You also want a slotted spoon for removing scum that forms on some fermenting foods. Choose spoons made of stainless steel or food-grade, heavy-duty plastic. Wooden spoons absorb odors and are hard to sanitize well.

Food processor

A food processor with a wheel for slicing and a grater attachment is very handy. It allows for more variety in the shapes and sizes of vegetables you can ferment, and it shortens prep time.

Hand grater

A more economical option than a food processor, a hand grater is a nice gadget to have. Buy one that's made well or it will break from vigorous use.

Yogurt maker

Yogurt makers are sold in kitchen appliance stores and can make the home production of yogurt easier.

Slow cooker or crock pot

You can use a slow cooker or crock pot when fermenting some yogurt recipes and other fermented products. A deep and large slow cooker would be best if you're going to buy one for fermenting.

Cheesecloth and/or coffee filters

Cheesecloth and/or coffee filters are invaluable both for filtering fermented foods and for covering containers that don't have a lid. You can hold them on with rubber bands or string. Cheesecloth allows gases and moisture caused by fermenting to escape while keeping out dust and other impurities. Cheesecloth is inexpensive, and you can usually find it in the canning section of stores.

Special equipment for meat fermenting

If you're interested in making sausage and other fermented meats, you need some additional equipment. Fermenting meat requires extra care when cleaning and sterilizing equipment to avoid food-borne illness, so if you have old, rusty equipment, you may want to purchase new.

Meat grinder

To make sausages and other forms of fermented meat, you need a meat grinder. Hand-crank and electric models of meat grinders are available. Hand-crank models are cheaper and suitable for most home meat processors.

Meat grinders are sized with numerals from 8 to 32. This size is the diameter of the *throat* — the area that you push the meat through — and basically determines how much meat you can grind at one time. A size 10 is good for smaller batches of meat.

Your meat grinder should have a variety of interchangeable plates that the meat is pushed through. The plates have holes in them that determine the size of the grind, from fine to coarse. Different types of meat and different recipes may call for different-sized plate holes. Also, you should be able to take your meat grinder apart so you can easily sanitize it.

Sausage stuffer

If you feel that meat fermentation and sausage-making is going to become a hobby, you may want to invest in a sausage-stuffing machine, which you can buy from specialty catalogs. These machines stuff ground meat into *casings* — natural or artificial skins that hold meat. Fermented sausages always need casings. Some meat grinders have attachments to stuff sausages.

When you try your first sausage recipe, you can stuff the sausages with a stainless-steel kitchen funnel. If you like the results, then invest in a sausage stuffer.

Room thermometer and hygrometer

When fermenting meat, you must control the temperature and humidity of the room it's fermenting in, and you can't do that unless you have instruments that measure them. A *hygrometer* measures humidity. We recommend a combination thermometer-hygrometer with a digital display. These are relatively inexpensive and can be found in many stores.

Special brewing equipment and tools

Unlike other methods of fermenting food, brewing requires some specialized equipment that you'll probably have to purchase. If you're lucky, you may be able to use a few pieces of equipment that you already have around the house. Some larger cities have stores that carry supplies for beer and wine making so you can examine and choose supplies locally. You can sometimes find used brewing equipment for sale. This section lists all the equipment that you'll need to begin brewing.

If this list sounds complicated, many places sell brewing kits that contain all the supplies a beginner needs. The salesperson in a brew shop may also be able to guide you in purchasing equipment if you explain what your goals are.

Brewpot with lid

You need a large, stainless-steel pot of at least 16-quart capacity to serve as a brewpot. One that holds 20 to 30 quarts is better. You use the brewpot in the first part of the brewing process to cook the grains. It needs a well-fitting lid.

Fermentation bucket with lid

The fermentation bucket, sometimes called the *primary fermenter,* is where you pour a cooled liquid after the first stage of brewing. You'll probably need to purchase this because it needs to hold at least 7 gallons, and it needs to have an airtight lid. The lid has an opening into which you insert a drilled rubber stopper and an airlock. These buckets are generally made of food-grade plastic.

Drilled rubber stopper

The rubber stopper fits into the hole in the fermentation bucket. It's called *drilled* because it has a hole in the center where you insert the airlock. Make sure you buy the right sized stopper to fit the hole in your fermentation bucket lid.

Airlock

The *airlock* is a specialized but inexpensive piece of equipment you need to buy. It allows carbon dioxide produced during fermentation to escape while keeping out air filled with harmful bacteria.

Brew spoon

You need a brew spoon for stirring and mixing. It should be made of stainless steel or food-grade plastic and have a handle that's 18 inches long or longer.

Hydrometer with cylinder

A *hydrometer* measures the density and alcohol content of your brew. It's made of glass and is quite sensitive and fragile, so you need to handle it carefully, especially after you imbibe your brew!

Food-grade plastic tubing

You need about 5 feet of plastic tubing for transferring your brew from one container to another.

Bottling bucket

A bottling bucket is usually made of food-grade plastic and has a removable spigot on the bottom. You attach the flexible tubing to the spigot and then to the bottling tube to fill your bottles during the last stage of brewing.

Bottling tube with spring valve

The bottling tube is a ridged piece of plastic with a valve on one end. You connect it to the flexible tubing running from the bottling bucket spigot. You insert it in a bottle and use the valve to control the fill.

Bottles

You must use glass bottles for bottling fermented products. Plastic bottles are hard to sterilize well and may give an off flavor. If you reuse bottles, they must be heavy glass without threaded tops. You can repurpose used beer and wine bottles, purchase used bottles from a commercial brewery, or purchase new bottles from a brew shop. For a 5-gallon batch of brew, you need about 40 16-ounce bottles, or the equivalent of 640 ounces.

New bottle caps

You need new caps for your filled bottles. You have to purchase these from a brewing supply store.

Bottle capper

The bottle capper is a device that seals the caps on the bottles. You need to purchase one from a brewing supply store. Many models are available, but one that attaches to a work surface with a vice is easiest to use.

Bottle brush

You need a bottle brush to clean out bottles. It should have soft bristles and fit through small bottlenecks.

Some additional equipment helpful in brewing

Many of these items may be in your kitchen already:

- **Small bowl or bowls:** You need bowls for various mixing tasks. You probably have some in the kitchen.
- **Funnels:** Funnels are helpful when pouring liquids. For brewing, they should be made of material that can be sterilized, such as stainless steel.

- ✔ **Strainers:** These should be made of stainless steel for brewing so that they can be sterilized.

- ✔ **Kitchen scale:** You need to weigh many products when brewing so that you're using the proper proportions of ingredients. Choose a scale that measures fractions of ounces. You can use a postal scale for most recipes. You can find scales in most stores.

Keeping Everything Clean

Because you put so much care and attention into the fermentation process, you don't want to neglect the most vital component: cleanliness. When you ferment foods, you want good bacteria to change the nutritional content of the food. You want to discourage bad bacteria from coming in contact with the food and spoiling it. Spoiled food can make you very ill or even cause death, not to mention that it's a waste of time and money. Without clean and sterile utensils, equipment, and containers, the fermentation process won't work, and you open up the foods to spoilage and unfriendly organisms.

Make sure that anything that comes in contact with fermenting food is kept clean. This includes countertops, food processors, measuring spoons, and jars and other vessels.

We must note that you can't always achieve a completely sterile environment; you can only do the best you can. Aim for a sanitary and clean environment and you're on the right track. When it comes to meat, dairy, and alcoholic fermentation, take extra precautions to keep tools and containers clean. These foods have a greater chance of attracting harmful microbes and causing foodborne illness. We cover these topics in more detail in Chapter 2.

Cleaning, sanitizing, and sterilizing: Three different and important procedures

When fermenting, having a clean work space and tools ensures that your good bacteria outnumber the bad. This means your chances of creating a delicious fermented food are much greater. In short, cleaning your work area and equipment is essential to the final product.

- ✔ **Cleaning** means washing containers and equipment with hot water and soap. You can do it in a dishwasher or by hand. You should remove all visible particles of food or dirt. Always clean things before sanitizing or sterilizing them.

✔ **Sanitizing** is done with chemicals. The easiest thing to use is chlorine bleach — bleach without scent or color added. Add 1 tablespoon of bleach to each quart of warm water and soak your containers and utensils in it for 5 minutes. Everything must be submerged in the water. You can use a clean sink or large tub for sanitizing. If you can't submerge an item, soak a clean cloth in bleach solution and wipe the item. Use a different cloth for each item.

Alternatively, brewing shops and stores that cater to commercial restaurants carry other chemical sanitizing agents. Follow the label directions for their use, and always rinse your containers and utensils with clean water after sanitizing.

You don't need to sanitize your containers and equipment if you're going to sterilize them.

✔ **Sterilizing** means boiling your equipment and supplies in water for 3 minutes. Your equipment and containers must be able to stand up to this temperature without melting. In some cases, you pour boiling water into larger containers and allow them to sit for a few minutes, but this isn't perfect sterilization. Some dishwashers actually have a sterilize setting that you can use.

Developing a cleaning work flow

Here are some general steps you should follow when preparing to ferment food:

1. **Wash all containers, utensils, and weights that you're going to use in a dishwasher or by hand with hot, soapy water just before use.**

2. **Sanitize or sterilize equipment and containers, as called for in the recipe.**

3. **Rinse items in cool, clean water (sterilized items don't need rinsing).**

4. **Air dry items or dry them with paper towels; use a fresh paper towel for each item.**

5. **Store items on clean paper towels on clean countertops or tables until you use them.**

6. **Remove pets and small children from the room before you begin to work.**

7. **Before beginning to work with food, tie back your hair if it's long, and scrub your hands, including under your fingernails.**

Don't use dishcloths or rags to dry the cleaned items. Cloth is notorious for holding huge quantities of harmful microbes, and you spread those from item to item as you wipe them. Instead, use a clean paper towel for each item if you need to dry them. Aprons made of cloth also spread bacteria, especially if you wipe your hands on them. Keep paper towels close by for wiping hands.

Getting Familiar with Common Ingredients

Each fermentation category has different food ingredients and requires different starters to begin fermentation. A starter contains some of the good bacteria you want in that food and helps get the fermentation off to a good start. In each chapter we tell you what ingredients and starters you need for each recipe. But some ingredients are common to most fermentation recipes, and we discuss them here.

Water

Water is the most important ingredient used in fermenting. It may seem like you can just turn on your tap and get clean water to use, but with fermenting foods, it isn't that easy. Many things in water can affect how well foods ferment and can also affect the taste. Municipal water treatment plants add chlorine or chloramine to water supplies to kill harmful bacteria, and these products can also kill the friendly microbes you want to attract. Though chlorine dissipates into the air if it sits a day or two and can be boiled out, these practices don't get rid of chloramine, which is the most popular water treatment these days.

To remove chloramines from tap water, it must pass through a good micro-filtration process, an activated carbon filter, or a UV light filter. Reverse osmosis and other filtration methods don't work. You can also use sodium thiosulfate and a product called *Campden tablets* to remove both chlorine and chloramines. You can buy these tablets at brewing and wine-making supply stores.

If you have access to well water, be aware that well water varies tremendously as to what it contains. You've probably had the well tested to see whether it has bacterial contamination, nitrates, or arsenic, but well water (or spring water) often has dissolved minerals like sulfur and iron that can affect the taste and quality of fermented foods without being a health hazard to those who drink it. If your well has a high mineral content (hard water) or a high salt content, two common well problems, the water may not be suitable to use for fermenting.

If you have a filter that removes most minerals, your well water may be suitable for fermenting, but water that is run through common water softeners may not produce the best water for fermenting because softeners often leave salt or other chemical traces in water. Soft water is alkaline, which is the opposite of the acidic conditions needed in fermentation, and the good microbes need to work harder to produce results. You can try using well water, especially if you like its taste, and see how things turn out. If you have a tap that isn't connected to a water softener, we suggest you draw water from that and run it through a charcoal filter for use in fermenting.

Probably the best water to use in home fermentation and brewing is distilled water. It's inexpensive and available almost everywhere. Unless you're making gallons of fermented produce or brewing large quantities of beer, this is a good choice for home fermenting.

When buying bottled water, make sure you get distilled water, not just water, often labeled *spring* water. Most bottled water is just someone's tap water and may contain chlorine and chloramines, just like your tap water.

Salt

Common table salt has iodine added, as well as some chemicals to keep it from clumping, and it isn't the best salt for fermenting. But most stores sell kosher, canning, and sea salt, and all those are fine for your fermenting projects.

When a recipe calls for salt, don't reduce or increase the amount unless the recipe tells you it's okay to do so. In fermenting, salt is often important for food safety, reduces bad microbes, and helps preserve food. Never use salt substitutes or reduced-sodium products in fermenting unless your recipe gives instructions for it. These products may prevent fermentation and compromise food safety.

A combination of salt and water is called *brine*. Brine is often used in fermenting recipes.

Sweeteners

Almost all fermenting recipes call for some sweetener, usually sugar. Sugar helps feed the good microbes as they get started fermenting your food. When a recipe calls for sugar, you can use organic raw sugar or regular table sugar.

When using white sugar, look for cane sugar specifically. If the package doesn't say cane sugar, it's beet root sugar, which is a genetically modified product.

If the recipe calls for honey as a sweetener, make sure you use pure honey — preferably local, raw honey. Store-bought honey is fine as a second choice, although recent studies have shown that up to 75 percent of store honey is no longer honey because it has been purified to remove the pollen that makes it special. If other sweeteners, such as agave syrup, are called for in your recipe, use them, because substituting other sweeteners may not give your fermented food recipe the intended results.

Don't use sugar substitutes or half sugar-half substitute products in a fermenting recipe. These don't feed the good microbes and may also contribute to an off flavor in food. After food is through fermenting, if you feel it needs additional sweetener to taste good, you can add a sugar substitute.

Sourcing the Best Foods for Fermentation

In this section we talk in general about the fresh foods you'll be using for fermentation. The different foods you can ferment include vegetables, fruits, nuts, seeds, grains, beans, meat, fish, and dairy. Whatever categories of food you're considering starting with, just know that working with whole, fresh foods is essential to fermentation.

Whole, organic, and local

Whole foods are foods that aren't processed. They are food in its natural form. In the fermenting process, starting with the freshest, highest-quality food you can get is important. This is not the time to use bruised peaches, overripe tomatoes, and old meat. It's especially important not to use any food that's already showing signs of mold or spoilage. The process of fermentation counts on good bacteria changing the food into something more. Damaged and old food may already have sizable colonies of bad bacteria on it, and these may overwhelm the good bacteria you want.

Choosing organic, local, and seasonal food ensures that the food is in its best state, with a better taste and more nutrients than food shipped from far away. Be sure to thoroughly wash all produce, even organically grown produce, before you use it. Both organic and conventionally farmed produce are sprayed with pesticides, herbicides, and other chemicals that leave residues that can affect the fermentation process and aren't good for anybody! In addition, all produce, even organic produce, can become contaminated with salmonella, E. coli, and other harmful organisms from growing conditions and handling during harvest. Wash all fresh produce!

Considerations for meat and dairy

For meat and dairy, you also want to choose locally produced food. Of course, when at all possible, choose meat and dairy from animals kept in humane conditions and raised without hormones and antibiotics. Fish should be wild-caught. The happier and more naturally the animal is raised, the healthier the animal and the better it is for you.

Beef, pork, lamb, and goat meat are the most common meats used in fermenting. Poultry is sometimes used, but the meat is soft and pale and doesn't always make a good finished product. Poultry also has a greater chance of carrying salmonella and other harmful disease organisms.

The most important considerations with meat and dairy products are that they're absolutely fresh and that they're kept chilled (below 40 degrees) from harvesting to your use of them. Just like with produce, meat and dairy products that are old or not properly stored already have a lot of bad microbes growing on them, which affects taste and food safety.

It should go without saying, but never eat or ferment meat from animals you find dead or that appear ill before they're slaughtered. Wild animals may have several other diseases that can harm humans, and fermentation isn't a safe way to prevent disease transmission. Venison from deer in tuberculosis-free areas is the best wild game to use for meat fermentation. Farmed buffalo and elk are generally pretty safe game meats also.

Organic food gets pesticides, too

Yes, even organic food is sprayed with pesticides, especially if the food is produced by a large, commercial organic farm. The pesticides used on an organic farm are organic; that is, they have a natural origin. However, if they kill pests, they may also affect you and the microbes fermenting your food. For example, *rotenone,* which was a common organic pesticide until just a few years ago, was found to cause several serious health problems in humans and isn't supposed to be used in food production anymore.

A small, local organic farm may use little or no pesticides on fruits and vegetables, and if pesticide use concerns you, you may want to purchase food from these growers. Just in case, however, wash all fruits and vegetables thoroughly, even those grown organically. Most pesticide residue is on the outside of the fruit or vegetable, and you can remove it with proper washing.

Part II
Vegetables, Fruits, Condiments, and Salsas

Visit www.dummies.com/extras/fermenting for tips on choosing the best quality foods to ferment.

In this part . . .

- ✔ From carrots to kimchi, find out which vegetables can be fermented and how.
- ✔ Select the right starters to use.
- ✔ Discover the world of alternative sweeteners.
- ✔ Find out how to ferment your own dips, spreads, and salsas.

Chapter 5

Vegetables

*N*othing's more rewarding than picking your own organic produce. In a perfect world, this would be from your own garden. Very few folks are lucky enough to do this. Otherwise, getting your produce from an organic delivery service (such as a CSA) or going to your local market or farmers' market is just as good. When you get to enjoy organic tomatoes, green beans, cabbage, and carrots in their prime, your palate is happy, and so is your body.

If you have abundance from your garden — and you likely will if you're growing at home — then the downfall is that you can only get through so many veggies and harvest fast enough before they start to go to waste or you run out of options of what to do with them. If you buy your produce elsewhere, you still may want to take advantage of those enzymes and nutrients when the veggie is freshest. That's where food preservation and fermenting your veggies comes in.

With some of the simple methods outlined in this chapter, you'll find out ways to preserve some of your favorite veggies. Of course, they won't only last longer, but you'll also be inspired to find new ways to enjoy them, depending on what they're preserved with.

Picking Produce for Fermenting

Choose fresh produce for your ferments. The key to fermenting is that the decay is controlled. You want only good bacteria to flourish — not the bad — so make sure your vegetables aren't past their prime. You don't want to introduce any unwanted bacteria in the equation.

Choose produce that's slightly under-ripe. Fermentation softens your veggies a bit, so don't start with anything overly ripe, or you may end up with mush!

Identifying ideal veggie choices

Naturally crisp and crunchy vegetables make satisfying ferments. You'll notice that many of the recipes in this chapter feature vegetables that come into season toward the end of summer and into fall. This allows you to create ferments that will last you well into the colder months.

- ✔ **Cabbage:** Use red or green cabbage for a simple sauerkraut or the Napa variety for a spicy kimchi. Cabbage is a reliable vegetable to use for fermenting. It softens and reduces in volume but doesn't lose its bite.

- ✔ **Cucumbers:** Nowadays, cucumbers are pickled with vinegar, salt, and other preservatives, but you can ferment cucumbers to make crispy, tangy fermented dill pickles.

- ✔ **Greens:** Kale, dandelion, and other greens are great additions to your vegetable ferments. They can be cultured on their own, but the strong taste may not please all palates!

- ✔ **Root vegetables:** Carrots, beets, parsnips, and other root vegetables are especially suited to fermenting. Their sweetness lends itself to fermenting, and they pair well with spices like cumin, horseradish, and dill.

Finding your garden delights

You can venture out to your local grocery store to buy vegetables for your ferments, but if you're looking for fresh in-season vegetables, there are a few other budget-friendly options you might try:

- ✔ **Community-supported agriculture (CSA):** These groups connect farmers with consumers who are willing to invest in the growing season. For a set fee, usually paid in installments, consumers receive a "share" of food each week. This setup supports farmers and ensures consumers have access to fresh food, farmed in a sustainable way. Some but not all CSAs are organic.

✓ **Local farmers' markets:** Most towns have a weekly market where local farmers share their freshest picks. Getting to know your farmers is a great way to source top-quality produce at the height of the season. If you have any questions about how the food is grown or if you're looking for a hard-to-find item, they'll be happy to chat!

✓ **Pick-your-own farms:** Some farms allow you to pick your own produce from their fields. Picking your own vegetables is a surefire way to guarantee freshness. And you can choose the size and quality of every item. It's cheaper too!

Sourcing the healthiest produce

If you want to make sure you're getting the highest quality organic and locally grown vegetables, here's what to do in order of priority. Make it work for you!

✓ Grow your own veggies at home.

✓ Become part of a local CSA; check online to find about them in your area.

✓ Shop at farmers' markets and smaller health food stores.

✓ Shop only in the organic section of your local grocery store.

The goal is to become an informed consumer, but more important, to give your body the nutrients it deserves from veggies. Then, when it comes time to ferment the veggies, they'll not only last longer but taste better!

Choose organic and GMO-free

Want the best flavors and health benefits from your recipes? When buying produce for your homemade recipes, do your best to choose organic or GMO-free. Certified organic produce helps you avoid putting chemicals and pesticides in the environment but also in your body. Similarly, looking for GMO-free labels and stickers ensures you know exactly what's in the food. These choices optimize the fresh flavor in your food. Try going organic or GMO-free for one day or one week — you'll be sure to notice the difference!

Selecting Starters for Vegetable Ferments

Vegetables contain all the lactic acid they need to trigger fermentation, so technically speaking, you don't need to add a starter. In practical terms, though, salt is almost always used for flavor and to slow fermentation. It also offers these benefits:

- Pulls water out of vegetables. This helps submerge them in liquid, inhibiting bacterial growth.

- Promotes lactic acid bacteria (the "good guys") so they can crowd out other bacteria.

The key to fermenting vegetables is to ensure that they're covered with a liquid. This creates an oxygen-deprived environment where "bad" bacteria is unable to grow.

If you'd rather not use salt — or would prefer to reduce the amount of salt you use in a ferment — Sandor Katz suggests these salt alternatives in his book *The Art of Fermentation:*

- **Caraway, celery, and dill seeds:** These are often added to vegetable ferments to kick up the flavor, but they also add some salt content to your batch.

- **Celery juice:** Juice a few stalks of celery and dilute with water to add to cover your vegetables.

- **Seaweed:** Kelp, arame, or hijiki can replace part of the salt in a ferment. Katz recommends rehydrating the seaweed and chopping and adding it to your vegetables, along with the soaking water.

Not all salt is made the same

You may be fooled into thinking that salt is salt! This can't be further from the truth.

Your only two good choices are

- **Sea salt:** This type of salt contains trace minerals that are important for your health, like sulfur, magnesium, boron, and silicon. Sea salt also has naturally occurring iodine. Sea salt is always slightly grey in color.

- **Himalayan pink salt:** This also is a natural salt choice containing many naturally

occurring minerals. It's slightly pink in color and has a profound effect on taste when used in food preparation and preserving.

Table salt is not an option! Table salt contains anti-caking additives to prevent clumping. It also has had iodine "added in" after it has been stripped out. Table salt is sometimes even bleached to keep it white. These unnaturally occurring compounds and processes can add an unpleasant taste to food and impede the fermentation process.

Mastering the Basics

Fermenting vegetables is an easy process that involves only a few steps and a bit of time.

Preparing your vegetables

You can chop, shred, or leave your veggies whole, depending on the vegetable you're using and the end result you want to achieve. Sauerkraut is most often shredded while cabbage for kimchi is traditionally chopped in larger pieces. Root vegetables can be cubed or sliced thin for a different texture. Cucumbers can be sliced into coins, halved, or left whole.

Salting

There are two basic approaches to salting vegetables for fermentation.

- ✔ **Brine:** Dissolve salt in water. Add to your fermentation vessel with the vegetables, making sure they are completely covered. Some recipes call for soaking vegetables in a brine for a few hours to soften them before draining and packing as usual.

- ✔ **Dry salt:** Sprinkle salt on your prepared vegetables, then massage and pound them (with a spoon or even your hands) until they release enough juice to cover them in your chosen fermentation vessel.

You can make a basic brine by dissolving 3 tablespoons of sea salt in 1 quart of water. Use this ratio to make whatever quantity of brine you need. Then just pour over your veggies and let the fermentation begin!

Packing the jars

Once your vegetables have been chopped and salted, it's time to pack them up. Add them to your jar or crock and press down to submerge the vegetables in their juices and release any air bubbles. This could take some time, so keep trying if at first it doesn't seem like there's enough juice to cover them. If the vegetables still aren't covered, you can add some filtered water.

See Chapter 4 for more information on choosing fermenting vessels.

Waiting and tasting (and waiting some more!)

How quickly they ferment will depend on a few factors, including how much salt was used in the preparation and how warm the location is. More salt will result in a slower ferment. A warmer room will speed it up.

Your vegetables will start to ferment after a few days, but the process will continue for much longer. Traditionally, fermented vegetables were prepared to preserve the harvest through winter, so keep that in mind when deciding when your batches are ready. The flavors will change and grow as days and weeks go by.

Once your ferment has reached its peak (according to your own tastes), move it to the fridge. It will still continue to ferment, but at a much slower pace. And it will probably be gone before too long!

Sauerkraut

Prep time: 30 min • **Ferment time:** 3–4 weeks • **Yield:** 4 quarts

Ingredients	Directions
5 pounds cabbage **3 tablespoons canning salt**	*1* Peel off damaged outer leaves of cabbage. Quarter the cabbage and remove the hard core. Finely shred the cabbage by hand or use a food processor.
	2 Scald a glass gallon jar with boiling water. Add a 1-inch layer of the cabbage mixture to the jar. Sprinkle with salt. Tamp firmly with a wooden utensil to remove any hidden air pockets and bruise the cabbage, making it release juice. Repeat Step 2 until all ingredients are used up.
	3 Cover the jar with a coffee filter and leave for 2 hours to allow time for the salt to draw out water from the cabbage. Every 30 minutes, tamp down the cabbage to help draw out brine and force it to be submerged.
	4 After 2 hours, if you still don't have enough natural brine, mix 1 teaspoon canning salt to 1 cup water and pour over mixture. When the cabbage is fully submerged, remove the coffee filter and place a small saucer that fits just inside the top of the jar so it rests directly on the submerged cabbage. Add a weight, such as a water-filled quart jar, to keep the saucer and product under the brine.
	5 Cover with a clean dishcloth to keep out dust and insects. Place the jar out of the way, at room temperature. Check the jar daily, for 3 to 4 weeks. Skim off any scum that may build up on the top. Replace the clean dishcloth with another each time you remove scum buildup.

Carrot and Caraway Sauerkraut

Prep time: 30 min • **Ferment time:** 3–4 weeks • **Yield:** 4 quarts

Ingredients	*Directions*
2 medium carrots 1 tablespoon caraway seeds 5 pounds cabbage 3 tablespoons canning salt	*1* Wash, peel, and grate the carrots with a box grater or a food processor equipped with a grating blade. Peel off damaged outer leaves of cabbage. Quarter the cabbage and remove the hard core. Finely shred the cabbage by hand or use a food processor. Mix together with carrots and add caraway.
	2 Scald a glass gallon jar with boiling water. Add a 1-inch layer of the cabbage and carrot mixture to the jar. Sprinkle with salt. Tamp firmly with a wooden utensil to remove any hidden air pockets and bruise the cabbage, making it release juice. Repeat Step 2 until all ingredients are used up.
	3 Cover the jar with a coffee filter and leave for 2 hours to allow time for the salt to draw out water from the ingredients. Every 30 minutes, tamp down the ingredients to help draw out brine and force the cabbage mixture to be submerged.
	4 After 2 hours, if you still don't have enough natural brine, mix 1 teaspoon canning salt to 1 cup water and pour over mixture. When the mixture is fully submerged, remove the coffee filter and place a small saucer that fits just inside the top of the jar so it rests directly on the submerged cabbage. Add a weight, such as a water-filled quart jar, to keep the saucer and product under the brine.
	5 Cover with a clean dishcloth to keep out dust and insects. Place the jar out of the way, at room temperature. Check the jar daily, for 3 to 4 weeks. Skim off any scum that may build up on the top. Replace the clean dishcloth with another each time you remove scum buildup.

Carrot Ginger Beet Sauerkraut

Prep time: 30 min • **Ferment time:** 3–4 weeks • **Yield:** 4 quarts

Ingredients	*Directions*
2 medium carrots	**1** Wash, peel, and grate the carrots and beet with a box grater or a food processor equipped with a grating blade. Peel ginger and grate using a microplane or mince with a sharp knife.
1 medium beet	
1 tablespoon ginger	
5 pounds cabbage	
3 tablespoons canning salt	**2** Peel off damaged outer leaves of cabbage. Quarter the cabbage and remove the hard core. Finely shred the cabbage by hand or use a food processor. Mix together with carrots, beet, and ginger.
	3 Scald a glass gallon jar with boiling water. Add a 1-inch layer of the cabbage, carrots, beet, and ginger mixture to the jar. Sprinkle with salt. Tamp firmly with a wooden utensil to remove any hidden air pockets and bruise the cabbage, making it release juice. Repeat Step 2 until all ingredients are used up.
	4 Cover the jar with a coffee filter and leave for 2 hours to allow time for the salt to draw out water from the ingredients. Every 30 minutes, tamp down the ingredients to help draw out brine and force the cabbage mixture to be submerged.
	5 After 2 hours, if you still don't have enough natural brine, mix 1 teaspoon canning salt to 1 cup water and pour over mixture. When the mixture is fully submerged, remove the coffee filter and place a small saucer that fits just inside the top of the jar so it rests directly on the submerged cabbage. Add a weight, such as a water-filled quart jar, to keep the saucer and product under the brine.
	6 Cover with a clean dishcloth to keep out dust and insects. Place the jar out of the way, at room temperature. Check the jar daily, for 3 to 4 weeks. Skim off any scum that may build up on the top. Replace the clean dishcloth with another each time you remove scum buildup.

Spinach Sauerkraut

Prep time: 30 min • **Ferment time:** 3–4 weeks • **Yield:** 4 quarts

Ingredients	Directions
1 cup chopped spinach 5 pounds cabbage 3 tablespoons canning salt	**1** Wash spinach in plenty of water, rinse well, and chop it into thin shreds.
	2 Peel off damaged outer leaves of cabbage. Quarter the cabbage and remove the hard core. Finely shred the cabbage by hand or use a food processor. Add spinach and mix together so they are evenly incorporated.
	3 Scald a glass gallon jar with boiling water. Add a 1-inch layer of the cabbage and spinach mixture to the jar. Sprinkle with salt. Tamp firmly with a wooden utensil to remove any hidden air pockets and bruise the cabbage, making it release juice. Repeat Step 2 until all ingredients are used up.
	4 Cover the jar with a coffee filter and leave for 2 hours to allow time for the salt to draw out water from the ingredients. Every 30 minutes, tamp down the ingredients to help draw out brine and force the cabbage mixture to be submerged.
	5 After 2 hours, if you still don't have enough natural brine, mix 1 teaspoon canning salt to 1 cup water and pour over mixture. When the mixture is fully submerged, remove the coffee filter and place a small saucer that fits just inside the top of the jar so it rests directly on the submerged cabbage. Add a weight, such as a water-filled quart jar, to keep the saucer and product under the brine.
	6 Cover with a clean dishcloth to keep out dust and insects. Place the jar out of the way, at room temperature. Check the jar daily, for 3 to 4 weeks. Skim off any scum that may build up on the top. Replace the clean dishcloth with another each time you remove scum buildup.

Kimchi

Prep time: 20 min, plus standing • **Ferment time:** 2–3 days • **Yield:** 1–2 quarts

Ingredients	*Directions*
1 medium Napa cabbage	***1*** Wash, core, and chop the cabbage into ½- to 1-inch pieces. Peel and shred the ginger and onion. Peel garlic, but leave the cloves whole.
1 to 2 inches fresh ginger	
1 medium onion	
1 head garlic	***2*** Place a layer of cabbage in a glass gallon jar. Sprinkle the salt over the cabbage. Repeat until all cabbage is used. Allow the cabbage and salt to sit at room temperature for 6 hours. Rinse the salt off the cabbage and place the cabbage in a large mixing bowl.
½ cup salt	
3 tablespoons soy sauce or tamari	
	3 Mix the ginger, onion, garlic, and soy sauce with the cabbage. Place the mixture in a glass gallon jar. Cover the filled jar with cheesecloth.
	4 Allow to ferment at room temperature for 2 to 3 days. Place the jar in the refrigerator to ferment for another week, or until the desired taste has developed.

Fermented Cucumbers

Prep time: 20 min • **Ferment time:** 3–4 weeks • **Yield:** 9 pints

Ingredients	*Directions*
4 pounds pickling cucumbers 2 quarts water ½ cup sea salt ½ cup dill 20 cloves peeled garlic	**1** Wash and trim the blossom ends of the cucumbers. Scald a glass gallon jar with boiling water. Combine the water with the salt, stirring to dissolve the salt.
	2 Place the dill and peeled garlic cloves in the bottom of a jar. Add the cucumbers to the jar. Pour the salt-water brine over the cucumbers. Weigh the cucumbers down with a saucer to keep them submerged.
	3 Cover the jar with a clean muslin cloth and place the jar in a cool place for 3 to 4 weeks. Check daily to see whether enough brine is covering the cucumbers, and change the cloth. After 3 to 4 weeks, place the jar in the refrigerator for storage.

Fermented Garden Vegetables

Prep time: 20 min • **Ferment time:** 1 week • **Yield:** 1 gallon

Ingredients	Directions
1 green pepper	*1* Scald a glass gallon jar with boiling water. Remove the seeds from all the peppers and thinly slice them into rings.
1 red pepper	
1 yellow pepper	
1 head garlic	*2* Peel the garlic cloves. Peel and thinly slice the onions. Cut the broccoli and cauliflower into florets.
2 medium onions	
1 head broccoli	*3* In a large pot, combine the salt and water. Place all the vegetables in a gallon jar and pour the saltwater brine over all. Place a saucer that fits inside the jar directly onto the vegetables to hold them under the brine.
1 head cauliflower	
2½ tablespoons sea salt	
2 quarts water	
	4 Cover the jar with a muslin cloth and place it in an out of the way place at room temperature. Check the jar daily, removing any scum, washing and replacing the saucer and recovering the jar with a clean cloth.
	5 After the fourth day, taste for a desired flavor. After 1 week, place the jar in the refrigerator for up to a month.

Vary It! Consider changing up your veggies by trying this with carrots, beets, green beans, or red cabbage.

Fermented Onions

Prep time: 10 min • **Ferment time:** 7 days • **Yield:** 1 quart

Ingredients	*Directions*
6 to 8 medium onions 3 tablespoons sea salt 1 quart water	**1** Peel and thinly slice the onions. Combine the salt and water. Place onions in a glass quart jar, pour the salted water over them and press down the onions so they stay under the brine.
	2 Cover the jar with a muslin cloth and place it in an out of the way location at room temperature. Check the jar daily, removing any visible scum and replacing the cloth. After 1 week, place the jar in the refrigerator to enjoy.

Fermented Dandelion Greens

Prep time: 20 min • **Ferment time:** 3–5 days • **Yield:** 1 quart

Ingredients	*Directions*
4 cups fresh dandelion leaves **1 tablespoon sea salt** **3 cups water**	*1* Wash the dandelion greens. Combine the salt and water, mixing until the salt is dissolved.
	2 Pack the dandelion greens into a quart jar, tamping firmly and pouring the salted water over the packed layer. Repeat the greens layers and the saltwater layer, ending with the greens fully submerged.
	3 Cover tightly and place at room temperature to ferment for 3 to 5 days. Place the fermented leaves in the refrigerator and enjoy.

Garlic Dilly Beans

Prep time: 10 min • **Ferment time:** 7 days • **Yield:** 1 quart

Ingredients	Directions
1 cup fresh green beans 3 cloves garlic	*1* Soak the green beans in cool water to get firm and crisp.
1 teaspoon dill seed or 2 fresh dill heads	*2* Pack all the ingredients in a 1-quart size jar.
Basic brine	*3* Pour over brine to cover all the beans.
	4 Leave 1 inch space at the top of jar, put on lid and band, and screw tightly.
	5 Allow to ferment at room temperature for 3 to 7 days.
	6 Transfer to the fridge when ready; they will keep for several months.

Note: This is the perfect way to make your tender sweet summer green beans last longer. They stay crisp in the brine and taste delicious next to some quinoa or in a salad.

Spiced Infused Pickled Beets

Prep time: 30 min • **Ferment time:** 3 days • **Yield:** 2 quarts

Ingredients	Directions
3 large beets	**1** Trim your beets, leaving the taproots and 2 inches of the stems. Wash and drain the beets using a stiff brush to remove any clinging soil. Peel your beets and slice them into thin large rounds.
2 cups thinly sliced white or yellow onions	
2 cloves garlic	**2** Prepare your jars and two-piece caps (lids and screw bands) according to the manufacturer's instructions. Keep the jars and lids hot. (See Chapter 4.)
1 teaspoon cloves	
3 cinnamon sticks, broken into 1-inch pieces	**3** Place the onions, water, garlic, cloves, cinnamon, and fennel seeds in the jar.
2 teaspoons fennel seeds	
3 cups salt water	**4** Add your beet slices into the jar, pour in the water and salt together, and make sure they are submerged in the brine.
1 tablespoon sea salt	
	5 Allow it to sit on the countertop for 3 to 4 days.
	6 Store in refrigerator for several months.

Pickled Asparagus

Prep time: 10 min • **Ferment time:** 3–4 days • **Yield:** 8 pints

Ingredients	Directions
12 pounds young asparagus spears	**1** Scald 2 to 4 jars in boiling water and set aside.
4 tablespoons dill seed	**2** Wash, trim, and break the asparagus into small pieces to fit into a jar.
4 teaspoons whole mustard seed	
8 cloves of garlic	**3** Put the dill seed, mustard seed, and garlic into your jars.
Basic brine: 10 cups water and ½ cup sea salt	**4** Put the asparagus stalks upright into each jar leaving top inch of the jar free. Put as many as you can into each jar.
	5 Pour the water and salt mixture over the asparagus enough to cover while leaving 1 inch free at the top of the jar.
	6 Leave to ferment for 3 to 4 days at room temperature.
	7 Transfer to cold storage.

Chapter 6

Fun with Fruits

*F*ermenting fruit is a fun way to enjoy fruit beyond its expiration date and one of the only ways to make fruit last without unnatural or artificial preservatives. The *lactobacilli,* which are "friendly" bacteria, are the primary agent that ferments the fruit.

Fruit is very versatile; you can use it in so many wonderful ways. In this chapter, you find fruit-filled recipes that are easy and fun to tackle. From jams and jellies to chutneys and preserves to fermented fruit leather, you have many different ways to explore the fermentation process using fresh fruit!

Fermenting Fruit for Long-Term Storage

Fermenting fruit is very different from fermenting vegetables. Fruit is made up predominantly of sugar, and the sugar affects the process. Fruit has a tendency to spoil quicker and turn alcoholic, as a result of the yeasts consuming the sugars. To avoid spoilage, you should ferment fruit for shorter periods of time and be sure to use a starter. We use whey, but extra salt is an alternative starter, although you need to add it carefully so that you don't overdo it and end up with an overly salty product. Savory recipes can tolerate more salt than sweet recipes.

Bacteria is your best friend

The lactobacilli that accomplish the process of fermentation are already abundant on most foods. However, the fermentation process encourages them to grow, multiply, and spread as they consume sugars in fruits and produce beneficial acids. This means that they're good for you!

When you keep fermented fruits in long-term storage, two likely scenarios take place:

- ✔ Nonalcoholic fruit ferments can become alcoholic during storage. Over time, the low acid levels and ready food sugar give the yeasts a chance to start working. As the yeasts consume the sugar, alcohol is created as a byproduct.

- ✔ Fruits are likely to spoil during long-term storage because the high sugar and low acids levels of the ferments attract spoiling organisms.

To avoid either of these situations, consume your fruit preserves relatively quickly, after a few weeks of fermenting them. However, some recipes, such as those with high acid levels, will last longer with lemons. Lemons have the ability to prolong the freshness of fruit because they reduce the decaying time. Any fruit that has a high acidity level will naturally extend the life of the ferment.

You should eat fruit recipes within a few weeks after they're finished. Plan to make small batches for you and your family and friends.

Selecting Ideal Fruits

Fermenting fruits is a great way to explore fruits you may not normally eat.

Here are several tips to keep in mind for your best fermentations:

- ✔ **Choose fresh fruits if possible.** Although you can create fruit fermentations using canned fruit, fresh fruits make controlling the levels of sugar in your recipes easier. Canned fruit is often a cheaper option, but your health and the recipe's flavor benefit from fresh fruit. You can also use frozen fruit, which is ideal for fruits that may not store well. Allow these fruits to thaw fully and then follow the fermented recipe as usual.

✔ **Select organic fruits when possible.** Organic foods are produced with no pesticides, fertilizers, or heavy chemicals, which are often linked to harmful consequences in both your body and the environment. If your fruits aren't organic, you should remove the skin whenever possible because most of the chemical residue is on the skin.

✔ **Explore your local options.** Depending on where you live and your seasons, local fresh fruit may be in generous supply. Why not grab a buddy and head to your local fresh fruit farmers' market or, heck, go out on a limb — climb a fruit tree in a friend's backyard or visit a pick-your-own fruit farm!

✔ **Look for fair trade.** Fair trade fruit certification ensures that farmers who grow your fruit are paid fairly. This means they're paid at least the minimum wage of the region in which they grow the fruit. Looking out for these labels on fruit supports a healthy economy.

✔ **Ensure that the quality and texture of your fruit matches your recipe.** One of the many secrets to great fermentations is selecting the correct quality of fruit to match your recipe. Is the fruit over-ripe or under-ripe? How will the ripeness affect your recipe? Some recipes may require you to have more natural *pectin,* a carbohydrate that's more abundantly found in less-ripened fruit. Jams and jellies, for example, call for more pectin, while other recipes may encourage you to select over-ripe fruits for better flavor.

✔ **Choosing imperfect fruits is okay.** It's okay to select fruit that may have a bruise or strange bump. You can always cut out the imperfection and no one will ever notice the difference. Depending on what recipe you're creating, you may very well need to mash or chop the fruit regardless.

✔ **Experiment with various fruit types.** Give yourself permission to play and have fun. So many different kinds of fruit are out there, from salmonberries to elderberries to breadfruit to dragon fruit. Test your local grocers for their varieties and see which flavors and textures suit you best.

✔ **Enjoy your fermented fruit recipes quickly.** As a general rule, lacto-fermented fruit recipes have a short shelf life. This isn't true for all fermented recipes, but most that include fruit. Keep this in mind.

A little bit about pectin

All fruits contain *pectin,* a carbohydrate that occurs naturally in fruits and is extracted and sold on the marketplace in both liquid and powdered form. Powdered pectin is commonly added in many jam and jelly recipes to make them more gelatinous. Pectin always exists in unripe fruit, but as the fruit ripens and matures, the amount of pectin decreases. Therefore, when choosing recipes that require more gelling, select fruit that has only just begun to ripen.

Nondairy Starters for Fruit Fermentation

A *starter culture* is a microbiological culture that actually performs fermentation. Starters usually consist of a cultivation medium — such as grains, seeds, or nutrient liquids — that have been well colonized by the microorganisms used for the fermentation.

Here are various options for some great, nondairy, fruit-fermentation starters:

- ✔ **Finished water kefir:** You use this at the same rate as whey: about ¼ cup per quart of ferment. (See Chapter 14 for more on kefir.)

- ✔ **Leftover fermented juice:** The juice of previously fermented pickles, sauerkraut, or other ferments is rich with beneficial organisms. You use it at the same rate as whey, a commonly used fermenting substance from milk. However, keep flavor matching in mind; a pickle juice starter probably won't taste very good inside fruit ferment.

- ✔ **Vegan probiotic starter:** You can purchase this easy-to-source powder and mix it with water or nondairy milk to make the equivalent of a whey starter.

- ✔ **Water kefir grains:** You use 1 tablespoon extra water kefir grains per quart of ferment and scale up from there. Similarly (although not nondairy), you can use 1½ teaspoons of extra dairy kefir grains per quart of ferment.

What if a recipe doesn't specify a starter culture and you want to use one? Generally, ¼ cup of liquid starter works for 1 quart of ferment. Scale up accordingly. If a recipe calls for another amount, by all means follow the wisdom of the recipe author. However, like many aspects of traditional cooking, starter culture usage isn't an exact science, and a range of amounts will probably work.

No whey, Jose!

Lactose intolerant? Interested in exploring a vegan diet? When using whey in fruit fermentation, recipes just say: No whey, Jose! *Whey* is derived from cheese and is a milk product. It may be familiar to you as a nutritional supplement; bodybuilders use it as an added protein source. If you do choose to explore using whey as a fermentation starter, keep in mind that the whey used as a nutritional supplement isn't the same type of whey used for fermentations, so you should explore making your own at home.

To Add Sugar or Not to Add Sugar? That Is the Question

You can use additional sugars in the fermentation process — or you can omit them. To begin with, you should strictly adhere to the required amount in a recipe. However, when choosing your sweetener of choice, make sure it's an unrefined sweetener.

White sugar versus alternative sugars

White sugar is the antithesis of what you want to put in your body, especially when you're trying to maximize the nutrition from your food. Although sugar is sugar, you can choose more healthful and natural options that actually carry with them some beneficial nutrients, minerals, and enzymes.

Try reducing sugar using natural alternatives or eliminating sugars altogether. Your body will thank you!

Here are some great sugar alternatives for your fruit-fermentation recipes:

- ✔ Coconut sugar or coconut syrup
- ✔ Honey, preferably raw
- ✔ Maple syrup or date syrup
- ✔ Unrefined cane sugar, rapadura, or Sucanat

How do I know what amount of sugar to replace?

Figuring out the amount of sugar to replace can often be a tricky area when it comes to substitutions and alternatives. The basic rule of thumb is to replace 1 cup of white granulated sugar with the equivalent dry version of granulated Sucanat, rapadura, or coconut sugar. However, when replacing white granulated sugar with a liquid form of sweetener, you can scale down by ¼ cup (use ¾ cup of liquid sweetener — such as honey, maple syrup, date syrup, or coconut nectar — per cup of white sugar).

Apple Cinnamon Chutney (Lacto-Fermentation)

Prep time: 15 min • **Ferment time:** 2–3 days • **Yield:** ½ gallon (2 quarts)

Ingredients	Directions
6 cups coarsely chopped apples	*1* Wash, quarter, and core the apples, and then coarsely chop them by hand or in a food processor.
½ cup lemon juice	
¼ cup vegan probiotic starter	*2* Combine all ingredients in a mixing bowl.
1 cup water	*3* Transfer the mixture to one clean half gallon jar, two quart jars, or a crock.
¼ cup rapadura, Sucanat, palm sugar, or other natural sweetener	
1 cup chopped pecans or other nuts	*4* Pack down so all ingredients are covered in liquid and are at least ½ inch below the rim of the container. Add more water if necessary to submerge all ingredients.
1 cup raisins or currants	
2 teaspoons sea salt	*5* Cover tightly. You can use a tight lid or securely fasten plastic wrap with a rubber band around the edge.
4 tablespoons five-spice blend	
	6 Let the container sit out at room temperature for 2 to 3 days.
	7 Check daily, or as necessary, for any mold growing on the surface and skim away, repacking carefully. Taste for desired texture.

Tip: If the weather is very hot, fermentation may only take a day or so. Burp the jar if necessary (to prevent explosions). When you're happy with the taste and texture, transfer the chutney to the refrigerator in an airtight container. This chutney keeps for a few weeks and should be eaten within a short time frame.

Vary It! Try using a variety of different apples, and explore other nuts such as walnuts or even pumpkin seeds.

Cranberry Walnut Chutney

Prep time: 5 min • **Ferment time:** 2–4 days • **Yield:** 1 quart

Ingredients	Directions
3 cups cranberries	*1* Mix the fruit and nuts together in a bowl. Add the salt, lemon juice, vegan starter, and ginger, and mix well.
½ **cup raisins**	
½ **cup walnuts, chopped**	
1 teaspoon sea salt	*2* Using a wooden spoon, pack the chutney tightly into jars. Pound down so that the fruit is quite compressed and the liquid rises. Add water as necessary to bring the liquid level with the fruit mixture, which should be about 1 inch below the top of the jar (you need to leave a little room for expansion during fermentation).
Juice of 1 or 2 lemons	
¼ **cup vegan starter**	
1 tablespoon (or more if you really like it) grated ginger	
½ **cup (or more) filtered water**	*3* Keep the chutney at room temperature for 2 to 4 days (open to check whether it's fizzy/bubbly, as this confirms fermentation) before transferring to the refrigerator.

Vary It! This recipe calls for 3 cups of fresh fruit; you could also try this with apples, plums, peaches, mango, pineapple, grapes, papaya, or a combination of fruit.

Peach Cinnamon Chutney

Prep time: 10 min • **Ferment time:** 3 days • **Yield:** 1 quart

Ingredients	Directions
1½ pounds (3 large) peaches, cut into quarters, pitted, and peeled (see tip below)	**1** Coarsely chop the peach quarters.
2 tablespoons Sucanat	**2** Place all the ingredients in a 1-quart Mason jar.
¼ cup vegan starter	**3** Be sure the liquid covers the peaches, and leave 1 inch of room at the top. If the liquid doesn't cover the peaches, add water to cover.
2 teaspoons sea salt	
1 teaspoon cumin seed	
1 teaspoon fennel seed	**4** Seal tightly.
1 teaspoon coriander seed	
½-inch piece peeled ginger root, minced (1 teaspoon)	**5** Let the chutney sit at room temperature for 3 days, or until the lid is taut, and then refrigerate unopened for up to 2 months. After opening, the peaches will keep for about 2 weeks.
½ cup raisins	

Tip: To peel peaches, drop the quarters into a pot of boiling water for 5 to 10 seconds. Remove them and plunge them into an ice-water bath. When the peaches are cool enough to handle, discard the skins.

Vary It! Try using cranberries or cherries as an alternative to raisins.

Date Raisin Chutney

Prep time: 10 min • **Ferment time:** 3 days • **Yield:** 2 cups

Ingredients	Directions
1 cup pitted dates	**1** Chop the dates and raisins.
1 cup raisins	
½ cup apple cider vinegar	**2** Place all the ingredients in a 1-quart Mason jar.
¼ cup Sucanat or coconut sugar	**3** Place the lid on and seal tightly.
¼ cup vegan starter	**4** Allow the chutney to sit for 3 days, until the lid is firm.
1½-inch piece of fresh ginger, peeled and minced	
½ teaspoon ground cumin	**5** Refrigerate and enjoy for up to 2 months.
1 cup water	

Tip: There's nothing like a sweet chutney to complement your spicy Indian meal. This chutney is great with a homemade spelt chapati as a dip or packed on some basmati rice. It's made from natural sugars that come from dried fruit and natural sweeteners.

Mango Orange Spread

Prep time: 10 min • **Ferment time:** 2 days • **Yield:** 2 quarts

Ingredients	Directions
6 cups dried mangoes, unsulfured	**1** In a medium bowl, combine the mangoes with warm water to cover. Let them sit until they become completely soft.
¾ teaspoon sea salt	
¼ cup raw honey	**2** Transfer the mangoes to a food processor. Add the salt, honey, orange rind, and vegan starter to the food processor. Blend together until smooth.
2 tablespoons orange rind	
¼ cup vegan starter	**3** Transfer the mixture equally between two wide-mouth quart jars, leaving enough room below the rim of the jars. Cover tightly and leave at room temperature for 2 days. (Trim off any mold daily and re-cover.)
	4 Transfer the jars to cold storage. This spread keeps for a few weeks.

Note: You want to use mangoes that haven't been preserved or treated with sulfites. Look for mangoes that aren't bright and shiny, which means they've ripened naturally, without any chemical preservatives.

Vary It! Replace 2 to 3 cups of the dried mangoes with apricots for a unique flavor.

Raspberry Mint Preserves

Prep time: 10 min • **Ferment time:** 2 days • **Yield:** 1 quart

Ingredients	*Directions*
6 cups fresh raspberries, washed and drained (or any other berry except strawberries, which are too acidic for this method)	**1** Mash all ingredients together until they're mixed and the berries are crushed.
30 mint leaves, chopped finely	**2** Put in a clean, quart-sized glass jar, leaving the top inch of the jar free.
⅓ cup sweetener of choice (recommended: coconut sugar or Sucanat)	**3** Cap tightly and ferment at room temperature for 2 days.
1½ teaspoons sea salt	
6 tablespoons vegan starter	
3 teaspoons powdered pectin	

Note: If any mold or scum appears at the top, skim it off. Transfer to cold storage (refrigerator or cellar) for up to 2 months or freeze for longer keeping.

Fermented Fruit Leather

Prep time: 15 min • **Ferment time:** 2–10 days

Ingredients

2–3 pounds of fruit (apples, bananas, pears, berries, mango, kiwi, or whatever you like)

1 teaspoon sea salt (optional)

Spices to taste (cinnamon, nutmeg, and so on)

Sweetener to taste (raw honey, maple syrup, agave, stevia, or sugar)

Half a packet of a vegan starter culture such as Caldwell's Starter Culture or Body Ecology Culture Starter (you'll need salt for the starter; the quantity depends on the specific culture you use)

Directions

1 Prepare the fruit (core the apples, remove unwanted seeds from other fruits, and so on) and cut it up into manageable pieces.

2 Process the fruit in a blender, food processor, or food mill until the mixture is relatively smooth.

3 Mix in the salt (if desired), spices, and sweetener. Prepare the starter culture of your choice according to the instructions that come with the culture prior to mixing it in.

4 Place the mixture in a canning jar (leaving 1 to 2 inches of headroom) and place a lid on the jar. Allow the jar to sit in a warm spot (70 to 80 degrees) for 2 to 10 days (2 to 3 days if using whey; starter culture instructions often recommend 7 to 10 days). When the culturing process is complete, use caution when removing the jar lid because pressure may have built up.

5 To dry the fruit leather, spread the fruit mixture on sheets of unbleached parchment paper, an alternative variety of plastic sheet that fits inside your food dehydrator, or a baking sheet if you're using your oven.

6 Allow the fruit leather to dry overnight or for 8 to 24 hours. The exact length of time depends on the temperature and the mixture's thickness. We recommend setting your oven or food dehydrator to 110 degrees or less to preserve the beneficial bacteria.

7 The fruit leather is done when it's smooth and no longer sticky.

8 Store the finished fruit leather in an airtight container.

Tip: If the mixture is very watery, try straining it through a tea towel or tight-weave cheesecloth to thicken it up a bit. The strained juice is delicious to drink and contains beneficial bacteria from the culturing process.

Preserved Lemons

Prep time: 30 min • **Ferment time:** 4 wks • **Yield:** 1 quart

Ingredients	Directions
3 lemons, scrubbed well	*1* Cut the lemons in quarters, stopping ¼ inch before the bottom so the wedges stay together.
3 tablespoons plus 3 tablespoons sea salt	*2* Sprinkle 1 tablespoon of salt over each lemon and rub it around in all the crevices.
1 bay leaf	
8 cardamom pods	*3* Pack the lemons inside a jar one by one, pressing down on each with a kraut pounder or meat hammer to pack in the jar well and release juices.
1½ teaspoons whole cloves	
1 teaspoon cinnamon chips	*4* Put the bay leaf in the jar. Crush the cardamom pods, cloves, and cinnamon in a mortar and pestle and add them to the jar.
Juice from 3 lemons	
	5 Mix the lemon juice with the rest of the salt. Pour it over the contents of the jar. You can add additional water if necessary so that the lemons are covered. Consider weighting the top of the lemons so they stay submerged. If the lemons are large enough and are packed well, weight may not be necessary.
	6 Cover the jar tightly and leave it at room temperature for 1 to 4 weeks. Burp the jar as necessary to release pent-up gases.

Tip: You'll know the lemons are done if the peel is soft and pliable — now quite edible. Also, the mixture may be bubbly. And the smell — wonderful, of course!

Chapter 7

Spreads, Dips, Condiments, and Salsas

In This Chapter

▶ Choosing healthy homemade condiments

▶ Getting creative with different flavors

▶ Varying things up with different vinegars

▶ Trying out recipes for dips, spreads, and condiments

Condiments can jazz up your meal with an intense burst of color and flavor. These add-ons are essential to great-tasting food and are a convenient way to add your personal touch to whatever you're making. With your own fermented condiments, you can control how sweet, tangy, or spicy your product is.

The recipes in this chapter are perfect for first-time fermenters because they're easy to make in small batches and to keep on hand. With fermented condiments, you don't need to invest a huge amount of time or effort or buy special equipment. Plus, they'll keep for months in the fridge or cold storage.

As we demonstrate in the recipes in this chapter, fermenting spreads and dips is an easy process. In many cases, you only need to add one extra step to your usual prep:

1. **Make your dip or spread and pack it into a jar.**

2. **Cover with a starter.**

3. **Cover with a lid and put in the cupboard for a few days.**

Reaping the Health Benefits of Homemade Condiments

In Chapter 3, we discuss how using fermented foods in every meal gives your body added probiotics and enzymes that help support healthy digestion. The good news is that you don't need a lot to get these benefits, which is where condiments, dips, and spreads come in. People love to dip and spread food; it's in their nature to smear this, scoop that, and add a dollop here and there. All you need is a couple of tablespoons of sauerkraut or guacamole (or both) to give your sandwich (and your body) a boost!

The difference between making your own condiments and buying condiments in stores is immense. That bright-red bottle of ketchup or salsa is probably loaded with sugar, fat, and chemical preservatives. Jarred and canned foods earn their long shelf life with chemical additives and highly refined ingredients. Save your money — and your health — by fermenting your own.

Experimenting with Flavor

Get creative in your ferments! Look in your pantry and choose your favorite flavor combinations to make your very own signature product, or concoct a dozen different variations to suit your mood. Here are some tried-and-true flavor options:

- ✔ **Hot:** Try different peppers like spicy cayenne, smoky chipotle, or hair-raising habanero for sauces that make you sweat.
- ✔ **Savory:** Turmeric, curry powder, cinnamon, and cloves can add some zing and depth to your condiments.
- ✔ **Sweet:** Honey, maple syrup, and coconut sugar add their own unique, sweet undertone.

When trying out new flavors, use a little at first to test your palate. A little bit of cinnamon or clove can go a long way. Or consider making a condiment plain the first time. You can push the envelope after you become familiar with a ferment's basic flavor profile.

Exploring Vinegars

Vinegar is a versatile condiment, and in its raw form, it's a raw food, alive with healthful bacteria. Use it on its own, in salad dressings, or to make other condiments, like ketchup or spicy chili sauce.

Everyone is familiar with distilled white vinegar — the plain, flavorless variety that's most suited to baking or pickling — but you can create vinegar from a number of different ferments. Here are some of the most common kinds of vinegar:

- ✔ Apple cider vinegar: This is the best one to use because it's the most active in terms of its nutritional value.
- ✔ Brown rice vinegar
- ✔ Red or white wine vinegar

The process

In some senses, vinegar begins when alcohol fermentation ends. During alcohol fermentation, yeasts feed on sugars. When all the sugar has been consumed, the all-you-can-eat buffet closes. But the process isn't over yet.

Acetobacter is a kind of bacteria that feeds on alcohol and turns it into acetic acid. After you expose alcohol to oxygen, these bacteria can set up shop in your alcohol solution and get to work. After a few weeks, your alcohol solution transforms into vinegar.

Ambitious do-it-yourselfers can start from the very beginning by turning a sweet solution into alcohol and then creating a vinegar. But the easiest and fastest way to make vinegar is to start with an already fermented alcohol like wine. Just add a *starter* of raw vinegar to help get things going, and cover with a piece of cheesecloth or a coffee filter.

It's all about your mother

A *mother* is the spongy mass that forms in unpasteurized vinegar. The mother takes a while to form in your fermenting vinegar, but after you have one, you can use it as a starter in your future vinegars.

If you don't want to make your mother from scratch, here are a couple of other ways you can get hold of one to turbo boost your vinegar:

- ✔ Buy a raw vinegar with the mother and add it to your solution.
- ✔ Share! Get to know other fermenting foodies and share your cultures.

Creamy Guacamole

Prep time: 10 min • **Ferment time:** 3 days • **Yield:** 8 servings

Ingredients	Directions
¼ cup fresh cilantro or parsley, chopped	*1* Chop and dice the cilantro or parsley, onions, and tomatoes and add them to a bowl.
¼ cup red onions, diced	
½ cup tomatoes, diced	*2* Quarter the avocados, remove the skins, and add them to the bowl.
3 ripe medium avocados	
1 lime, juiced	*3* Add the lime juice to the bowl. Mash the avocados with a fork and mix all the ingredients together.
1 teaspoon sea salt	
¼ cup brine	*4* Season the mix with salt to taste.
	5 Add the brine to the mix. Transfer the guacamole to another jar if fermenting. Cover the surface tightly with plastic wrap and leave on the kitchen counter for up to 3 days.
	6 When you see bubbles forming under the plastic wrap, taste the guacamole. When it has the tanginess you like, store it in the refrigerator.
	7 When it's ready to serve, skim off the top layer or mix it into the dip. This guacamole keeps in the fridge for 2 weeks.

Basil Garlic Pesto

Prep time: 10 min • **Ferment time:** 3 days • **Yield:** 8 servings

Ingredients	Directions
2 cups fresh basil leaves	**1** Wash the basil. Remove the stems and just tear off the leaves.
¼ cup olive oil	
2 cloves garlic	**2** Place the basil in a food processor on the pulse setting to get it roughly chopped and packed down.
2 tablespoons lemon juice	
1 teaspoon miso paste	**3** Add in the olive oil, garlic, lemon juice, miso paste, honey (if desired), nuts, and sea salt. Blend everything until the pesto is finely chopped and well combined.
1 tablespoon honey (optional)	
¼ cup pine nuts or walnuts	
2 tablespoons sea salt	
2 tablespoons brine	**4** In a large Mason jar with a wide mouth, pack the pesto in tightly and pour the brine over the top, just covering the pesto.
	5 Seal the lid on tightly and let the pesto ferment in a dark place for 3 days, and then place in your fridge.

Tip: Top your favorite gluten-free or whole-grain pasta with this recipe for a nourishing meal. Don't forget to add some extra greens.

Hot and Spicy Chili Sauce

Prep time: 15 min • **Ferment time:** 3–5 days • **Yield:** 1 pint

Ingredients	Directions
16 fresh jalapeño, sriracha, or other hot peppers, with stems removed	*1* Place the peppers, garlic, salt, and brine in a food processor. Pulse to chop and then run the processor until the mix forms into a paste.
2 cloves garlic, crushed	
1 teaspoon sea salt	*2* Transfer the mix to a wide-mouth pint jar. Cover the jar with a lid or air lock.
2 tablespoons brine	
	3 Leave at room temperature for 3 to 5 days, until bubbles form.
	4 Put a fine mesh sieve over a small bowl. Spoon the mix into the sieve. With the back of a spoon, press the mix against the sieve to release the hot chili sauce.
	5 Transfer the sauce to a bottle. Store the sauce in the refrigerator; it will keep for many months.

Miso Tahini Spread

Prep time: 10 min • **Ferment time:** 3 days • **Yield:** 8 servings

Ingredients	Directions
1 tablespoon miso 2 tablespoons tahini	**1** Place the ingredients in a medium bowl and mix together for desired consistency.
1 teaspoon tamari ½ lemon, juiced	**2** Add water (if desired) to thin out.
2 tablespoons water (optional to thin out spread)	**3** Allow the mixture to ferment for up to 3 days.

Note: Serve over steamed veggies or use as a salad dressing.

Vary It! For a variation, add in 1 teaspoon grated ginger and 2 tablespoons shredded carrot.

Homemade Ketchup

Prep time: 5 min • **Ferment time:** 3–5 days • **Yield:** 1 pint

Ingredients	*Directions*
2 cups tomato paste	**1** Spoon the tomato paste into a large mixing bowl and fold in the honey (or your natural sweetener of choice).
¼ cup raw honey (or another natural sweetener)	
¼ cup brine + 2 tablespoons brine	**2** Whisk in ¼ cup brine, apple cider vinegar, sea salt, allspice, and cloves. Continue to whisk until the paste is smooth and uniform.
2 tablespoons raw apple cider vinegar	
1 teaspoon sea salt	**3** Spoon the ketchup into a Mason jar. Top with the remaining 2 tablespoons of brine, cover loosely with a cloth or lid, and allow the ketchup to sit at room temperature for 3 to 5 days.
1 teaspoon ground allspice	
½ teaspoon ground cloves	
	4 After 3 to 5 days, uncover the ketchup and give it a thorough stir before transferring it to the refrigerator.

Honey Garlic Dijon Mustard

Prep time: 5 min • **Ferment time:** 3–5 days • **Yield:** 1 pint

Ingredients	Directions
1 cup yellow mustard seeds	**1** Soak the mustard seeds overnight. Drain and grind them into a paste.
½ cup filtered water	
1 tablespoon basic brine	**2** Mix the mustard paste with the rest of the ingredients in a small bowl.
1 teaspoon turmeric powder	
3 cloves garlic	**3** Place the mustard in a jar and cover tightly with a lid.
3 tablespoons raw honey	
1 tablespoon virgin olive oil	**4** Leave at room temperature for 3 to 5 days and then refrigerate.
2 teaspoons sea salt	

Garden Salsa

Prep time: 10 min • **Ferment time:** 3 days • **Yield:** 2 quarts

Ingredients	*Directions*
1 large onion or large bunch of green onions, diced	*1* Combine the onion, bell peppers, garlic, and cilantro in a food processor. Pulse three to five times until coarsely chopped.
3 small bell peppers, cored and diced	
6 large garlic cloves, peeled and minced	*2* Add the tomatoes, a third at a time, pulsing between additions to make room for them all to fit.
½ cup packed cilantro leaves, chopped fine	
2½ pounds roma tomatoes, diced	*3* Pour the pureed tomatoes into a large bowl. Add the lemon juice, sea salt, cayenne powder, and brine. Stir well and allow to sit a few minutes.
1 lemon, juiced	
3 tablespoons coarse sea salt	*4* Wash two quart jars or one ½-gallon jar, a food funnel, and jar lid(s) well with soap and hot water.
¼–½ teaspoon cayenne powder	
½ cup brine	*5* Ladle the salsa into the jar(s), leaving 2 to 3 inches of head space. Add water to submerge the salsa.
¼–½ cup water	
	6 Close the lid(s) tightly and leave the salsa at room temperature for a few days until it's bubbly and fermented.
	7 During this process, the solid vegetables may separate from the liquid. Simply stir with a wooden or plastic spoon until redistributed and submerged under the liquid.
	8 Transfer to the refrigerator to store.

Tomatillo Salsa Verde

Prep time: 20 min • **Ferment time:** 4 days • **Yield:** 2 pints

Ingredients	Directions
2 pounds tomatillos, husks removed, cored	*1* Dice the tomatillos into ¼-inch pieces and combine with all the other ingredients in a large bowl.
1 large yellow onion, diced	*2* With a separate clean spoon, taste the mixture. It should taste overly salty (salt is important in the preservation process and prevents the salsa from going bad while the fermentation process kicks in; the salty taste will dissipate after fermentation). If it doesn't taste overly salty, add another tablespoon of salt and mix to incorporate.
4 large Anaheim chilies, seeded and stems removed	
4 garlic cloves	
2 tablespoons finely chopped fresh cilantro	
2 teaspoons ground cumin	*3* Spoon the mixture into a large, clean jar and pour any remaining juices in as well.
½ teaspoon sea salt	
½ teaspoon crushed red pepper flakes	*4* Press the mixture into the jar so it's fairly compact. If the salsa isn't submerged in the natural juices, add a few tablespoons of water until it's covered.
¼ cup fresh lime juice	
	5 Put a secure lid on the jar and set it on your counter (out of sunlight) for 4 days.

Creating salsas that sing

Nothing is fresher than a raw salsa prepared at the height of summer. It's a taste we wish we could enjoy all year long. Bottled pasteurized salsas can seem bland and overcooked. To kill unwanted bacteria, pasteurized salsa is heated and packed into hot, sterilized jars to be boiled. Though this process is necessary from a food-safety perspective, it affects the salsa's taste, texture, and nutritional content.

The good news is that fermented salsas offer the best of both worlds. You can savor the freshness of summer vegetables and enjoy the benefits of eating a living food.

Part III
Grains, Seeds, Nuts, and Beans

Visit www.dummies.com/extras/fermenting for more information on the nutritional benefits of plants, seeds, and nuts.

In this part . . .

- ✔ Discover the difference between grains and pseudo-grains.
- ✔ Understand the benefits of soaking, sprouting, and fermenting your grains for optimal nutrition.
- ✔ Make your very own sourdough starter from scratch.
- ✔ Find out how to prepare beans for fermenting.
- ✔ See the healing and nutritional benefits of miso and tempeh.
- ✔ Get introduced to the variety of nuts, seeds, and tubers that can be fermented.

Chapter 8

Grains

In this chapter we explore the world of grains. So many different types of grains are available from every country and culture. You want to familiarize yourself with these various grains to discover how you can ferment them in a variety of recipes.

Getting to Know Your Grains

Your body needs grains as an essential source of complex carbohydrates and fiber. Whole grains are slow burning, which means they provide your body with long-lasting energy.

If your definition of eating whole grains is eating a whole-wheat bagel, then think again. Eating whole grains means actually eating the whole grain in its entirety and not just the flour. That means searching for flour and flour products that include every part of the seed and grain in the final product.

If the grain has been processed, crushed, or cooked, it may not deliver the same benefits. Look for whole grains that haven't been processed to get the best balance and nutrients from each grain seed.

Ancient grains

An *ancient grain* is a grain, usually derived from a grass, that has its roots in history. These grains have been used in countries and cultures for thousands of years and are said to be full of added nutrients that you don't find in the average grain from your grocery store.

Table 8-1 lists top ancient grains to include in your diet.

Table 8-1	Ancient Grains		
Name of Grain	**Health Benefits**	**How to Use It**	**How to Store It**
Oats: A category that includes oats, oat bran, and oatmeal	Oats contain fiber, which helps lower cholesterol levels. Oats are also beneficial for stabilizing blood-sugar levels and contain antioxidants, which reduce the risk of cardiovascular disease.	For all types, it's best to add oats to cold water and then simmer for 10 to 30 minutes, depending on the variety. For raw porridge, soak rolled oats overnight for a ready-to-go breakfast in the morning.	Store oatmeal in an airtight container in a cool, dry, dark place. It will keep for approximately two months.
Barley: A rich and chewy grain	Barley is an incredible source of soluble fiber and B vitamins. Both are essential for lowering cholesterol and protecting against heart disease.	After rinsing, add 1 part barley to 3½ parts boiling water or broth. After the liquid returns to a boil, turn down the heat, cover, and simmer.	Store barley in a tightly covered glass container in a cool, dry place. You can also store it in the refrigerator.
Kamut: A type of ancient wheat that's high in protein	Kamut has 20 to 40 percent more protein than wheat and is richer in several vitamins and minerals, such as calcium and selenium. Unlike some grains, kamut has a low oxidation level and retains most of its nutritional value after grinding and processing.	Soak 1 cup kamut overnight. Then add 3 cups of water and bring it to a boil, add a pinch of salt (if needed), turn the heat to low, and simmer for 40 to 45 minutes, or until tender.	Store kamut in a tightly sealed glass container in a cool, dry place.

Name of Grain	Health Benefits	How to Use It	How to Store It
Spelt: An ancient grain with a nutty flavor	In addition to being an amazing source of fiber and niacin, spelt is also easier to digest than wheat.	After rinsing, soak spelt in water for 8 hours or overnight. Drain, rinse, and then add 3 parts water to 1 part spelt berries. Bring to a boil, then turn down the heat and simmer for about an hour.	Store spelt grains in an airtight container in a cool, dry, dark place. Keep spelt flour in the refrigerator to best preserve its nutritional value.
Brown rice: A grain that's more wholesome than white rice; it's the unhulled and unmilled version	Brown rice is light, gluten-free, and an incredible source of long-lasting carbohydrates for energy. It's also high in fiber and contains several vitamins and minerals, as well as protein.	After rinsing brown rice, add 1 part rice to 2 parts boiling water or broth. After the liquid returns to a boil, turn down the heat, cover, and simmer for about 45 minutes.	Because brown rice still features an oil-rich germ, it's more susceptible to becoming rancid than white rice and therefore should be stored in the refrigerator. Stored in an airtight container, brown rice will keep fresh for about six months.

Pseudo-grains (seeds)

Pseudo-grains are more seed than grain, but they often get lumped into the grain category. They're seeds that are very high in protein, fiber, and low-glycemic carbohydrates. They include quinoa, buckwheat, teff, amaranth, and wild rice.

Pseudo-grains have a few things in common:

- They're all actually derived from seeds or grasses.
- They contain essential vitamins and minerals like calcium, iron, and magnesium.
- They're a source of slow-burning energy.

The other incredible feature about these grains is that they're all alkaline-forming and gluten-free, making them accessible to everyone. They're easy to digest, absorb, and assimilate. They make for incredible side dishes, cereals, loafs, muffins, and pilafs.

Table 8-2 lists the top pseudo-grains to include in your diet.

Table 8-2	Pseudo-Grains		
Name of Grain	**Health Benefits**	**How to Use It**	**How to Store It**
Amaranth: A seed of a plant from Central America that has a nutty flavor and can be combined well with other grains	Higher in protein than many other grains, amaranth also contains the essential amino acid lysine, which is typically hard to find in plant-based foods. It's a good source of calcium and iron, important for bone health. It's also high in potassium, phosphorous, and vitamins A, E, and C, and it's filled with essential oils that help lower cholesterol and blood pressure.	Use 1 part seeds to 2½ parts water. Bring to a boil and then reduce the heat to a simmer.	Keep amaranth fresh in a tight-fitting container with a lid. It's best stored in a cool, dry, dark place.
Buckwheat: A fruit seed that's related to rhubarb and sorrel, making it a suitable grain substitute for people who are sensitive to wheat or other grains that contain protein glutens	Buckwheat is rich in *flavonoids,* which are phytonutrients that protect against disease by extending the action of vitamin C and acting as antioxidants. It's a great source of protein, manganese, and vitamins B and E. Buckwheat helps balance and lower cholesterol levels while also protecting against heart disease. It also has mood-enhancing and mental-clarity properties.	After rinsing, add 1 part buckwheat to 2 parts boiling water or broth. After the liquid returns to a boil, turn down the heat, cover, and simmer for about 30 minutes.	Place buckwheat in an airtight container and store it in a cool, dry place. Always store buckwheat flour in the refrigerator; keep other buckwheat products refrigerated if you live in a warm climate.

Name of Grain	Health Benefits	How to Use It	How to Store It
Quinoa: A seed that has a fluffy, creamy, slightly crunchy texture and a some-what nutty flavor when cooked	Quinoa is a complete protein, providing all essential amino acids. It's also high in fiber, calcium, and iron.	Add 1 part quinoa to 2 parts liquid in a saucepan. After bringing the mixture to a boil, reduce the heat to a simmer and cover. One cup of quinoa cooked this way usually takes 15 minutes to prepare.	Store quinoa in an airtight container. It will keep for a longer period of time, approximately three to six months, if you store it in the refrigerator.
Teff: A grain that appears purple, gray, red, or yellow-ish brown; the seeds range from dark reddish brown to yellowish brown to ivory	Teff leads all the grains by a wide margin in its calcium content, with a cup of cooked teff offer-ing 123 milligrams. It's also an excellent source of vitamin C, a nutrient not commonly found in grains. It's a source of dietary fiber that can benefit blood-sugar management, weight control, and colon health.	Cook teff for about 20 minutes, with 1 cup of teff in 3 cups of water.	Store teff seed or flour on your shelf for up to one year in a dark, cool place, sealed in a container.
Wild rice: An aquatic seed found mostly in the upper fresh-water lakes of Canada, Michigan, Minnesota, and Wisconsin; when cooked it has a nutty flavor	Wild rice is a source of *lysine* (an essential protein) and B vita-mins. It has almost twice the protein content and almost six times the amount of folic acid as brown rice.	Put the grains into a saucepan with warm water to cover, and stir the rice around to allow any particles to float to the top. Skim off the particles and drain the water. It's best to repeat the rinsing one more time before cooking. Use 1 cup of dry wild rice to 3 cups of water. Cover and bring to a boil over high heat. Turn the heat down to medium-low and steam for 45 minutes to 1 hour.	Seal wild rice in a dark glass or opaque container in a cool, dark place for up to three years.

Soaking and Sprouting Grains

Sprouting any beans or grains begins to break them down and make them more digestible. Soaking grains signals them to begin their own digestion and break down some hard-to-digest proteins before they even hit your stomach. Sprouting ultimately makes it a lot easier for your body to digest and absorb the grains' nutrients (think less gas and more nutrients). The *phytic acids,* or anti-nutrients that actually prevent healthy minerals from being absorbed in your body, also become neutralized, and the enzyme inhibitors are released.

To sprout a grain, you must soak it for at least eight hours. Then you need to rinse and drain the grain (similar to a bean or a seed) for one to three days before a sprout grows.

After you soak your grain, it's more easily edible. The most beneficial part of a sprouted grain is that its nutrition profile doubles. Also, many gluten-containing grains, such as kamut and spelt, become much easier for someone who is gluten-sensitive to consume. But note that if you have celiac disease, you need to avoid all forms of gluten.

After a grain is soaked you can blend it to make batters, breads, and muffins. You can cook grains for porridges and pilafs, and of course you can sprout and ferment them. You have plenty of exciting recipes to explore, or you can start by just making your own sprouts!

Infamous Sourdough and Its Starter

When you begin exploring the world of fermentations, you'll hear people talking about sourdough, sourdough starters, and wild yeasts. A sourdough starter is really just another name for any fermented grain. It's a grain that starts to predigest during the fermentation process, which ultimately makes the flavor more sour.

Here are just a few great reasons to make your very own sourdough:

- ✔ Sourdough is delicious.
- ✔ Sourdough helps improve your digestion.
- ✔ Sourdough predigests gluten and phytic acid.
- ✔ Sourdough saves you the cost of buying commercial yeasts.
- ✔ Making the sourdough starter and keeping it alive is fun.

You can use anything from spelt flour to whole wheat to kamut to other ancient grains to experiment with sourdough. You'll be surprised just how many recipes you can make, from muffins to pancakes to breads . . . anything goes!

Feeding your sourdough starter with tender loving care

If you care about your sourdough starter, you'll feed it. Just like your body, sourdough requires food. The fermentation process feeds off of the sugars and carbohydrates in the flour. To keep the sourdough alive, consistent and regular feeding of flour is required. As long as you keep feeding your sourdough and store it in a good environment, it will stay alive, and you can keep making more recipes with it.

Sometimes you'll even hear people call the sourdough starter the *mother* because it creates other sourdough babies. Sourdough starter is very different from commercial yeasts, which require you to continue buying them and have a strong need for a stable environment. Sourdough is unique because it predigests the gluten and phytic acids in the grain.

Choosing the type of sourdough starter to use

All sourdoughs require a sourdough starter, the essential life force that makes your amazing breads and other recipes. The interesting thing about these starters is that they take on the flavor of the naturally occurring yeast and bacteria in your environment and the flour you use. From one batch to the next, each sourdough is a bit unique and different from the last!

You can purchase dry sourdough starters or find a friend who has a wet sourdough starter. Starting your own is possible but can take up to one week.

A dry sourdough starter may come as a dehydrated formula that contains sourdough yeast. (We suggest you look out for more traditional starters rather than manufactured yeasts for better flavor.) A wet sourdough starter is a more traditional culture that has been started from scratch, fed continuously with flour, and is now bubbling and alive. It's actually wet and sticky. When such a wet starter is fully thriving, it should be fed with water and flour at least once a week. See the upcoming recipes for a sourdough starter recipe, along with recipes for breads, muffins, and more.

Sourdough Starter

Prep time: 5 min • **Ferment time:** 5–6 days • **Yield:** 2 cups

Ingredients	*Directions*
2 cups flour of your choice (see tip after recipe directions)	**1** Place ¼ cup of water into a wide-mouth jar. Then add ⅜ cup of your chosen flour and mix well.
2 cups filtered water	**2** Stir the flour and water well. This brings in air and helps oxygenate your starter. Make sure you scrape all the sides and get all the flour into the starter.
1 packet dry yeast	**3** Cover the jar with a cloth napkin or paper towel and secure with a rubber band.
	4 Let the starter sit in a relatively warm place for 12 to 24 hours. Room temperature is ideal; you don't want to exceed 85 degrees.
	5 By now you may see some bubbles in your mixture. Repeat step 1 process every 12 to 24 hours in the same jar as your starter begins to expand. Your starter will start to get bubbly and fluffy and have a slight sour smell.
	6 If you don't see any bubbles yet, your starter may need help to make room for good bacteria. Try removing about half of the starter from the jar and adding another ¼ cup of water and ⅜ cup of flour. Let the starter sit for another 12 hours until bubbles occur and you see it grow in size.

Tip: Whole-wheat flour is a good choice to start, but we encourage you to play with spelt or kamut, too. Remember that the flour you choose affects the density and flavor of your recipe. Whatever you do, don't use white flour.

Note: If you're using a dehydrated starter, simply open the package, empty the contents into the jar, and follow the same steps.

Note: The starter will get bigger the more you feed it. You may need to discard some of it or transfer it to larger containers along the way. Also, the presence of liquid (which you can stir in) and a sour smell are normal.

Traditional Sourdough Bread

Prep time: 10 min • **Cook time:** 45 min • **Yield:** 1 large loaf

Ingredients	Directions
2 cups kamut sourdough starter	*1* In a large mixing bowl combine the starter, coconut sugar, and salt. Adding 1 cup of flour at a time, stir in as much as you can with a wooden spoon.
1 tablespoon coconut sugar	*2* When the dough is firm enough to knead by hand, roll it on a lightly floured surface until the dough is no longer sticky. Knead it by hand so it is moist and flexible!
4 cups kamut flour	*3* Wash the mixing bowl with some hot water, dry it, and place the dough in the warm bowl. Cover the bowl with a dishcloth and let the dough rise in a warm place until it has doubled.
2 teaspoons salt	*4* Now for the fun part: Punch down the dough! Form it into one round loaf and let it rise a second time on a lightly greased cookie sheet.
	5 Bake the loaf in a preheated oven at 350 degrees for 40 to 50 minutes or until the crust is brown. The bread should feel hollow when you give it a tap!
	6 It's hard to resist, but let your bread cool before you slice and serve it!

Vegan Sourdough Pancakes

Prep time: 10 min • **Cook time:** 15–20 min • **Yield:** 15 pancakes

Ingredients	Directions
1 cup spelt sourdough starter	**1** Place your sourdough starter into a medium mixing bowl.
1 cup spelt flour	
1 tablespoon ground flax + 3 tablespoons water as an egg replacement	**2** Add the remaining ingredients and mix very well.
2¼ cups rice milk	**3** Using about ½ cup of batter for each pancake, cook them on a hot, greased griddle until golden brown.
3 tablespoons coconut oil	
2 tablespoons coconut sugar	**4** Serve while hot with some delicious fresh fruit or soy yogurt.
1 teaspoon vanilla extract	

Spelt English Muffins

Prep time: 20 min, plus rising time • **Cook time:** 7–8 min • **Yield:** 12–16 pancakes

Ingredients	Directions
4 cups spelt flour	**1** Mix together the spelt flour, cornmeal, baking soda, and salt.
7 tablespoons organic cornmeal	
	2 In a separate bowl mix your wet ingredients: your sourdough starter and your vegan soured milk.
1 teaspoon baking soda	
¼ teaspoon salt	**3** Mix these ingredients until the dough is firm.
2 cups spelt sourdough starter	
	4 Knead the dough on a lightly floured surface until it is moist and flexible but not sticky!
¾ cup vegan soured milk (¾ cup rice milk with 1 tablespoon apple cider vinegar)	
	5 Roll the dough into ½-inch thickness. Cover and let it rise for 10 minutes. Cut out the muffins by using the round edges of a water glass, or find a great round cookie cutter.
	6 Dip the muffins into the remaining cornmeal and place them on a nonstick cookie sheet. Let this rise for 45 minutes to one hour.
	7 On medium-high heat, cook the muffins on a dry griddle. Cook them 3 or 4 at a time and turn them very carefully. Each muffin should be cooked about 7 to 8 minutes.

Note: Ye ol' English muffin is a classic morning food. Have one with a side of eggs or just plain with some butter (or coconut oil with a sprinkle of salt). This version is great for vegans or those seeking a healthier alternative.

Oaty Spelt Bread

Prep time: 30 min • **Cook time:** 30 min • **Yield:** 2 loaves

Ingredients	Directions
1 cup quick-cooking oats	*1* Place the oats, ½ cup spelt flour, coconut sugar, salt, and coconut oil in a large mixing bowl.
½ cup + 5 cups spelt flour	
½ cup coconut sugar or sucanat	*2* Dissolve the yeast in the warm water. Add this to the batter.
1 tablespoon salt	
2 tablespoons coconut oil	*3* Add spelt flour to your batter until it is smooth and elastic dough. Knead for 5 minutes.
1 package dry yeast	
½ cup warm water	*4* Place the dough in a greased bowl, cover with a towel, and put in a warm place to let rise for 1 hour. Your dough will soon double.
	5 For 30 minutes, bake at 350 degrees in the oven. Remove the bread and place on a cooling rack. Wait until cooled to eat.

Rye Spelt Bread

Prep time: 20 min, plus rise time • **Cook time:** 45 min • **Yield:** 2 loaves

Ingredients	*Directions*
1½ **cups warm water**	*1* In a large mixing bowl, combine the warm water and yeast. Allow to sit and bubble for 5 minutes.
4 **teaspoons yeast**	
¾ **cup sucanat or coconut sugar**	*2* Add coconut sugar or sucanat, molasses, salt, caraway seed, rye flour, and coconut oil and mix it well. Add spelt flour until the dough forms a ball. Roll the ball on a lightly floured surface and knead for 10 to 15 minutes until smooth and elastic.
¼ **cup molasses**	
1 **tablespoon salt**	
2 **tablespoons caraway seed**	
2½ **cups rye flour**	*3* Place the dough in a greased mixing bowl, cover, and allow to rise for 1 hour.
2 **tablespoons coconut oil**	
3 **cups light spelt flour**	*4* Punch down your dough and let it rise and double! This should take about 45 minutes.
	5 After your second rising, divide the ball of dough into two pieces. Shape into two loafs and place in two greased pans.
	6 Let these rise again for 1 hour. Yes, the third time is the charm.
	7 Preheat your oven to 375 degrees and bake your loaves for 45 minutes.
	8 Remove and place on a cooling rack. Let cool for 45 minutes to 1 hour before eating.

Spelt Sourdough Chapati

Prep time: 30 min • **Ferment time:** 12–24 hr • **Cook time:** 10 min • **Yield:** 10 servings

Ingredients	Directions
2 cups spelt flour (substitute 1 cup for 1 to 2 cups chickpea flour) ½ teaspoon salt ½ cup warm water 1½ tablespoons coconut oil	**1** In a large bowl, combine flour and salt. Add water and mix together with hands. Knead with your hands to form a firm, smooth dough. If too sticky, add more flour, 1 tablespoon at a time; if too dry, add more water, 1 tablespoon at a time. **2** Cover and let sit for 12 to 24 hours. The dough will begin to rise. **3** Pour melted coconut oil over dough and knead again. **4** Divide dough into 10 evenly sized balls. On a lightly floured surface with a rolling pin, roll out each ball evenly into a 7-inch circle. **5** In a dry, heavy-bottomed or nonstick frying pan on medium heat, cook each chapati by placing it top-side down in the pan. When it begins to bubble, after about 1 minute, gently press edges with a nonmetal spatula to allow it to gather air or rise. **6** Flip and cook the other side for 1 minute. Set aside in an ovenproof dish.

7 Repeat with remaining dough. If the pan gets too hot, reduce the heat. Experiment to determine the optimal heat to cook them.

8 Stack the chapatis on top of each other in the dish, separated with wax paper or coated with a little coconut oil to prevent sticking.

9 Cover the dish with a lid to prevent drying and place in a warm oven until ready to serve.

Note: Chapati, or unleavened bread, is a staple in India and a great addition to any meal. It makes great wraps for sandwiches or sides to rice and veggies.

Vary It! This chapati is made with spelt, a form of wheat that's high in protein and fiber. You can make your chapati completely gluten-free by using brown rice, chickpea, or buckwheat flour.

Rustic Fermented Corn Bread

Prep time: 15 min • **Ferment time:** 12–24 hr • **Cook time:** 20 min • **Yield:** 6–8 muffins

Ingredients	Directions
1½ cups yellow cornmeal	**1** Soak cornmeal and flour in water, lemon juice, and yogurt. Allow to sit at room temperature in a bowl covered with a cloth in a spot away from light for 12 to 24 hours.
½ cup teff flour or brown rice flour	
1 cup filtered water	
1 tablespoon lemon juice	**2** Preheat oven to 350 degrees.
¼ cup coconut yogurt	**3** In a large bowl, stir together cornmeal mixture, baking powder, and salt.
1½ teaspoon baking powder	
½ cup applesauce	**4** Add the applesauce, milk, maple syrup, oil, vinegar, and corn (if desired).
¾ cup plain rice milk	
¼ cup maple syrup	**5** Mix together gently until "just" mixed.
2 tablespoons grapeseed oil	
½ teaspoon apple cider vinegar	**6** Spoon mixture into lightly oiled muffin tins, filling cups to the top.
1 cup corn kernels (optional)	
	7 Bake for 15 to 20 minutes.
	8 Test with a knife to see if done. Cool for 5 minutes on a wire rack before serving.

Delicious Banana Bread

Prep time: 10 min • **Ferment time:** 12–24 hr • **Cook time:** 20–40 min • **Yield:** 8–12 servings

Ingredients	Directions
2 cups organic spelt flour (divided)	*1* Place 1 cup spelt flour in a bowl and cover with water. Allow it to sit for 12 to 24 hours.
⅓ cup applesauce + 1 tablespoon ground flax (if vegan)	*2* Preheat oven to 350 degrees.
2–3 ripe bananas, mashed	*3* In a small bowl, combine the ground flax with apple-sauce until it thickens.
½ cup rice milk	
¼ cup coconut oil, melted	*4* In a large bowl, place the mashed bananas, flaxseed mixture, rice milk, and coconut oil.
¼ cup maple syrup	
1 teaspoon baking soda	*5* In another bowl, combine the remaining 1 cup of spelt flour, maple syrup, baking soda, baking powder, cinnamon, and sea salt.
1 teaspoon baking powder	
¼ teaspoon cinnamon	
½ teaspoon sea salt	*6* Then mix wet ingredients into dry and fold slowly, and mix together until there are no lumps. Stir in blueberries or chocolate chips.
1 cup blueberries or chocolate chips	
	7 Drop by spoonfuls into an oiled muffin tray or pour batter into a loaf pan.
	8 Bake for 20 minutes if these are muffins or 30 to 40 minutes if this is a loaf.

Tip: Using spelt flour instead of whole-wheat flour gives these muffins a bit more depth in flavor and loads them with fiber and protein. Also, adding ripe bananas (the ones with brown speckles on them) gives this bread natural sweetness and makes it moist, tender, and delicious!

Chewy Chocolate Brownies

Prep time: 15 min • **Ferment time:** 12–24 hr • **Cook time:** 30 min • **Yield:** 10 servings

Ingredients	Directions
1 cup oat flour	**1** Oil an 8-×-8-inch baking pan and set aside.
½ cup unsweetened cocoa powder	**2** Measure oat flour into a large bowl, add the cocoa, baking powder, and salt and whisk together.
1½ teaspoons baking powder	
¼ teaspoon salt	**3** In a separate bowl, mix together sucanat, applesauce, and oil. Add the rice milk and vanilla extract and blend well.
1 cup sucanat	
½ cup applesauce/apple butter or soaked dates	
⅓ cup grapeseed oil	**4** Add the wet ingredients to the dry, using a rubber spatula to mix the ingredients until just blended. Stir in the chips.
½ cup vanilla rice milk	
1½ teaspoons vanilla extract	**5** Pour the batter into the pan and smooth the top.
⅓ cup dark chocolate chips	
	6 Bake for 35 minutes or until tester in the center comes out clean. Let cool completely before cutting.

Apricot Amaranth Muffins

Prep time: 15 min • **Ferment time:** 12–24 hr • **Cook time:** 25 min • **Yield:** 10 muffins

Ingredients	*Directions*
1 cup brown rice flour or teff flour	*1* Sift together flours in a large bowl. Add cinnamon, baking soda, baking powder, and salt and whisk to mix well. Add water or rice milk and apple cider vinegar and stir, forming a thick dough.
½ cup amaranth flour	
¼ cup arrowroot flour	
¼ teaspoon cinnamon	*2* Cover and let sit for 12 to 24 hours at room temperature.
1 teaspoon baking soda	
1 teaspoon baking powder	*3* Preheat the oven to 350 degrees.
¼ teaspoon sea salt	
½ cup water or rice milk	*4* Boil the water. Then place the hot water with one-half of the apricots and the ground flax or chia in a blender until a paste is formed.
1 teaspoon apple cider vinegar	
¼ cup hot water	
1 tablespoon ground flax or chia	*5* Melt the coconut oil to liquid and in a small bowl mix together the oil, maple syrup, and apricot and chia paste along with the pieces of apricots.
½ cup chopped apricots (for blending)	
¼ cup coconut oil	*6* Combine the wet mixture with the dough and form a solid batter and then add in the puffed amaranth.
¼ cup maple syrup	
½ cup chopped apricots (left as whole pieces)	*7* Scoop batter into a muffin pan and place in the oven. Bake at 350 degrees for 25 minutes until a toothpick comes out clean.
¼ cup puffed amaranth	

Note: Amaranth is an ancient grain that lends new and exciting flavors to your baking. These delicious muffins strike a nice balance with the nutty flavor of amaranth and the sweet tang of apricots.

Tip: Vegans can substitute 1 tablespoon ground flax or chia for 1 egg in any recipe. This helps bind the baked item while giving it a boost of fiber.

Chapter 9

Beans

Choosing to ferment beans takes your beans to a whole new level. It adds flavor and nutrition to a really simple and very inexpensive food.

Fermenting beans makes them easier to digest and helps your body absorb their nutrients. And for vegetarians and vegans, fermented beans add a wonderful source of protein and allow for some diversity in the diet.

Some forms of fermented beans offer diverse textures and tastes that pair really well with other staples in a plant-based diet, including veggies, grains, nuts, and seeds.

Beans, Beans, the Musical Fruit

Not all beans can be sprouted, but any kind of bean can be fermented.

The process is the same for fermenting cooked beans as sprouted beans. You need to add some brine, and the final product will taste different depending on the bean you select, such as

✔ Black beans

✔ Black-eyed peas

✔ Garbanzo beans (also called chickpeas)

✔ Kidney beans

- Navy beans
- Pinto beans
- Soybeans

Buying and storing beans

When you buy beans, do your best to make sure you're getting them as fresh as possible. We know this is hard to tell when they're dry. But if a bean has been sitting around in a bag or a bulk bin too long, it won't ferment properly, and it may have an off taste. So if you buy beans in bulk, find out how often the store changes them over, and if you buy beans in a package, find brands that you know and trust.

The best way to store dried beans is in glass jars in the cupboard. Try not to keep more than a jar full at a time. Buy them fresh as needed.

Preparing beans

You can prepare beans in many ways. That's what makes them so versatile and wonderful to use in a variety of recipes. They taste different with each method you use.

For the record, you need to soak all beans, regardless of what you plan to do with them. After you soak them, the next step is to cook, sprout, or ferment them, depending on which process you desire.

Cooking beans

After you soak your beans for at least 8 to 10 hours, you need to drain and rinse them. When you're ready to cook the beans, place them in a pot, cover them with fresh water, and add a piece of kombu.

Kombu is a sea vegetable that allows beans to cook faster and absorb more flavor. It also reduces the bloating some people get from beans.

Bring the water to a boil, and then reduce the heat and allow the beans to simmer for approximately 1 to 3 hours, depending on the bean variety. You can find many bean cooking charts online.

From here, you can ferment the cooked beans or even cook them again depending on the recipe.

Sprouting beans

After you soak beans for 8 to 10 hours and rinse and drain them, you can sprout them by placing them in a damp, aerated spot for the next 8 hours. The best choices are a fine mesh colander, a jar with cheesecloth or a perforated lid, or a big open bowl. After 8 hours, you need to drain and rinse the beans again for the next 1 to 3 days, depending on the bean variety.

The best and (easiest) beans to sprout are mung beans, chickpeas, and lentils.

You can buy sprouting kits in most health food stores, but it's just as easy to sprout at home using three readily available household objects: a jar, cheesecloth, and a rubber band. Here are the basic steps to sprouting:

1. **Rinse the beans well and pour them into a jar (fill about one-fourth of the jar).**
2. **Fill the jar with water.**
3. **Soak the beans overnight at room temperature.**
4. **The next day, pour out the water and beans and rinse with fresh water.**
5. **Return the beans to the jar.**
6. **Cover the jar with cheesecloth and secure the cloth with a rubber band.**
7. **Briefly turn the jar upside down to drain the remaining water.**
8. **Sprouts will begin to appear within 24 hours (give or take); they're ready when the shoot or sprout is as big or bigger than the item itself.**
9. **Rinse your sprouts before eating, and store them in the fridge for up to a week in a sealed container.**

Fermenting beans

Whether beans are cooked or sprouted, they can be fermented. Fermentation escalates the nutritional value of the bean in terms of vitamins, enzymes, and probiotics, and it also makes the beans easier to digest, last longer, and taste better.

When it comes to fermenting beans, you can use many of the methods discussed in this book. Either a basic brine recipe (water and salt) or a vegan yogurt starter (which you can purchase online) will work to activate the fermentation process for your bean recipe. (See Chapter 1 for some basic brine tips.)

Soy and Fermented Soy Foods

Soy is a species of legumes, widely grown for its edible bean, which has numerous uses. Soy has traditionally been used in Asian and Japanese cultures as a condiment. Soy is added to many commercial products — including milk, cheese, and other packaged products — as a stabilizer or enhancer. Soy is loaded with protein, fat, and immune-enhancing properties. Nowadays, people include soy as the main part of their meal.

As a result of the health craze regarding soy foods being good, consumers are purchasing anything and everything containing soy, thinking they're on the path to health. The issue with this overconsumption of soy-based products is that the soy used in the commercial industry is nonorganic, genetically modified, and so fractionated from its original form that it's no longer a food. Soy isn't bad for you, but you need to choose the right kinds of soy and eat it in moderation.

Here's a list of the top fermented forms of soy that you can include in your diet. These are the types of soy that are traditionally used in ancient cultures. You typically enjoy them in moderation and use them as a condiment, garnish, or accent to a meal. And, guess what, some of these you can even make at home!

- **Miso:** A fermented soybean paste with a salty, almond butter–like texture. Make miso soup or put miso in a salad dressing or marinade.

- **Natto:** Fermented soybeans with a sticky texture and a strong, cheese-like flavor.

- **Tamari or nama shoyu:** Traditionally made by fermenting soybeans, salt, and enzymes, tamari is the modern, healthy version of soy sauce. It is pure and has great flavor. You can also get it as low-sodium and wheat-free. It's great in salad dressings, sauces, and marinades.

- **Tempeh:** A fermented soybean cake with a firm texture and a nutty, mushroom-like flavor. Enjoy tempeh in a stir-fry, on sandwiches, ground up into "burgers," or just as is! See the tempeh recipes later in the chapter to make your own tempeh at home.

Comparing tofu and tempeh

Even though tofu and tempeh both come from the soybean, they're very different. This section looks at the differences between them and the uses for both.

Marni's delicious knowledge: What is GMO-free and why choose it?

Do you know exactly how your food is made? Do you want the best health benefits from your recipes? Buying certified GMO-free produce — produce that has no genetically modified organisms — helps you avoid any unwanted changes that happen to your food before it's placed on shelves. Sometimes, companies enhance or change food to extend its shelf life, freezing point, color, texture, or other factors that make it more marketable. Some of these changes are still being tested, and the long-term effects aren't well known. Do some research and decide for yourself whether GMO-free food is the right choice for you. When buying produce for your homemade recipes, I suggest you do your best to choose GMO-free to gain the purest food choices for your body's needs.

Tofu

Tofu is a processed soybean product that has not been fermented. It is also referred to as *bean curd*. Most people use tofu as a replacement for meat because it's high in protein and easy to cook with. You can grill it or fry it or eat it raw, in salads or as a side with rice. Be sure to consume tofu in moderation, and make sure it's organic and GMO-free when possible. Although tofu is a healthy way to enjoy soy, it doesn't have the same level of protein, enzymes, and probiotics as tempeh, which is a fermented product. (Because tofu isn't a fermented product, we don't include it in any recipes in this book.)

Tempeh

Tempeh can seem strange at first, but you'll quickly learn to love it. It's a fermented soybean on which grows an edible mold called *Rhizopus oligosporus*. The mold is grown in careful conditions that bind the tempeh and create an environment for good digestion. Tempeh makes the soybean far easier to digest and brings in an element of healthy living bacteria for the gut.

Tempeh requires a powdered mold as a starter. You can purchase this starter from well-researched online sources or from your local health food shop. You can leave tempeh starters sealed in your freezer until you're ready to use them. In addition to a starter and your soybeans, you need the following items to make tempeh:

- **Plastic bags:** You place these bags over your soybeans and poke small holes in them for breathability. Use a new bag for every new tempeh batch to avoid mold cross-contamination.

- **Thermometer:** Use a proper thermometer for food that can measure the temperature during the fermentation process.

Making the most of miso

Miso is not the easiest recipe to make, but it adds a nice, salty flavor to rice or vegetables. Miso is typically store-bought, but you can make it at home (see the first recipe in the following list of recipes). You can make it from red beans, black beans, or white beans, with a low- or high-salt content. Miso requires a lot of patience to make and time to ferment — up to several years for the true miso connoisseur. But hey, the best things in life are worth waiting for!

Tantalizing Tempeh

Prep time: 1 hr 15 min • **Cook time:** 2 hr • **Ferment time:** 36–48 hr • **Yield:** 1 quart

Ingredients	Directions
2 cups dry soybeans	*1* In a medium 6-quart pot, put the soybeans on medium-high heat and bring to a boil for 15 minutes. Leave the lid on during the boiling process.
2 tablespoons apple cider vinegar	*2* Remove the beans from the stovetop and let them sit in the boiled water for 2 hours. This helps to loosen the bean hulls.
¾ teaspoon tempeh starter	*3* Drain the soybeans and then place them back into the pan. Mash them well to loosen the hulls even further.
	4 Cover the beans in fresh water. Remove the bean hulls as they float to the top of the water. Gently boil the beans without the hulls for 30 minutes, or until tender. Remove, drain, and pat dry with a kitchen towel (or allow them to be spread and to air dry).
	5 In a clean bowl, pour in the mashed beans. Add the cider vinegar and tempeh starter. Spread the bean mixture thinly on the inside of two quart-sized plastic bags. Poke needle-size holes in the bags for ventilation.
	6 Place the bags into a dehydrator with an accurate thermometer at 85 to 95 degrees for 12 hours, or until you see the tempeh begin to heat up. This happens naturally from the fermentation process.
	7 Keep an eye on your tempeh over a 24-hour time period. You'll notice white mold growing on the surface; this mold is a good sign that the fermentation is working! Leave your tempeh in a dehydrator for up to 30 hours.
	8 When the tempeh is fully covered in this white, fizzy mold, it's ready to be steamed. This stops the fermentation process. Steam the tempeh for 25 minutes and it's complete. Refrigerate it until you're ready to use it. You can even remove some of the mold from the tempeh before you eat it, but you don't have to!

Miso

Prep time: 12 hr • **Cook time:** 40 min • **Ferment time:** 6–10 mo • **Yield:** 6 cups

Ingredients	*Directions*
3½ cups dry soybeans	*1* Soak the soybeans. Let them rest under water overnight in a large bowl or pot.
500 grams rice koji (see Chapter 14 for a koji recipe or purchase koji at your local Japanese health food store)	*2* Simmer the soybeans and then drain and rinse them. They should be enlarged from soaking overnight. Cover them again in fresh water in a large pot and simmer on the stove for 2 to 4 hours, or until tender.
1 cup sea salt	*3* Let the soybeans cool down. Remove 1⅔ cups of water from the simmered soybeans and place this on the side. Drain the soybeans and let the water and beans cool to room temperature.
	4 In a large pot, add the koji culture to your beans. The koji is responsible for breaking down the carbohydrates and sugars in the beans. Add in the 1⅔ cups of cooled soybean water and the sea salt. Stir well.
	5 Cover the bottom of a crock pot (½ gallon should do it) with a light layer of sea salt (about 1 teaspoon). Place your beans into the pot and pack them well with a wooden spoon. Add a sprinkle of salt on top.
	6 Cover the beans with parchment paper. Place a heavy plate that fits inside your pot on top as a weight to hold down the beans. Cover the whole thing with a cloth and tie it down with a rubber band if you can.

7 Let your covered beans sit in a cool, dark location for up to 6 months. Check on your fermenting beans at first every couple of days and then slowly every month. You should notice liquid begin to rise to the top. This is a good sign that your fermentation is creating soy sauce! Give it a good stir and make sure you have enough weight on the beans to extract the liquid. You'll see your beans darken and get more red tones over time, and you can taste them along the way to see when they're ready.

Tip: White mold is a normal change that may occur on the surface of your miso. Just scrape it off and keep on doing what you're doing! If you ever see black or blue mold, throw it out and start the entire recipe over. The warmer the environment, the faster the fermentation takes place, and the more air the miso is exposed to, the more likely mold will come to the surface.

Bringing-Home-the-Bacon Tempeh

Prep time: 20 min • **Marinate time:** 4 hr • **Cook time:** 10 min • **Yield:** 4 servings

Ingredients	Directions
8 ounces tempeh	*1* Slice the tempeh into ¼-inch-thick slices.
5 tablespoons tamari	
1 tablespoon honey	*2* Combine the tamari, honey, garlic, and liquid smoke and place the mixture in a zippered plastic bag.
1 clove garlic, minced	
1 tablespoon liquid smoke	*3* Place the sliced tempeh in the bag of marinade for 4 hours, turning the tempeh over after the first 2 hours.
5 tablespoons oil for frying	
Dash black pepper	*4* Fry the tempeh in oil until crispy, about 10 minutes.
	5 Sprinkle some black pepper on top for flavor!

Vary It! Want to try some new flavors? Vegans use nutritional yeast for some added B12 vitamins. This is delicious yeast found at health-food shops that you can fry up with your tempeh to give it an added cheesy flavor.

Tempeh Stir-Fry

Prep time: 25 min • **Marinate time:** 10 min • **Cook time:** 10 min • **Yield:** 10 servings

Ingredients	Directions
4 tablespoons toasted sesame oil	**1** To make the sauce, combine the sesame oil, tamari, garlic, ginger, orange juice, and syrup in a blender and blend until smooth, or mix in a bowl until well combined.
½ cup tamari	
4 cloves garlic, minced	
2 tablespoons chopped ginger	**2** Pour half the sauce over the tempeh or tofu to marinate.
Juice of 2 oranges	
2 tablespoons honey or maple syrup	**3** Heat some grapeseed oil in a skillet and add the tempeh or tofu. Stir-fry until crispy on the outside, about 10 minutes. Remove from pan.
1 package of organic tempeh or sprouted tofu	
Grapeseed oil	
1 white onion, sliced	**4** Place the veggies in the pan in order: onions, carrots, celery, bok choy, and broccoli.
2 carrots, cut diagonally	
2 celery ribs, cut diagonally	**5** Stir the veggies quickly on high heat, add in the remainder of the sauce, and add back in the tempeh or tofu.
2 heads bok choy, chopped	
1 head broccoli, cut into florets	
1–2 cups cooked brown rice	**6** Pour over cooked brown rice and garnish with white or black sesame seeds.
2 tablespoons black or white sesame seeds	

Napa Cabbage Salad with Sweet Miso Dressing

Prep time: 10 min • **Yield:** 6–8 servings

Ingredients	Directions
1 whole napa cabbage, washed and sliced into shreds	**1** Place the cabbage in a large salad bowl with the edamame (if desired). In a smaller bowl, make the dressing by combining the miso, honey nectar, mirin, sesame oil, lemon juice, and ginger and mix well.
1 cup shelled cooked edamame (optional)	
½ cup white miso	
⅓ cup honey nectar	**2** In a blender, blend all the dressing ingredients until smooth.
½ cup mirin (a tasty Japanese wine vinegar)	
¼ cup sesame oil	**3** Pour the dressing over the salad mixture in the bowl, and then top with the seeds and chopped nori or dulse and toss. Serve immediately.
¼ cup lemon juice	
¼ cup chopped ginger	
⅓ cup toasted sunflower seeds	
2 tablespoons sesame seeds	
1 sheet of nori or dulse, chopped	

Note: Nori and dulse are forms of seaweed. These ingredients add some exceptional flavor to the salad giving a nice natural dose of salt and minerals.

Orange Maple Ginger Marinade

Prep time: 10 min • **Marinate time:** 4–12 hr • **Cook time:** 10–30 min • **Yield:** 4–6 servings

Ingredients	Directions

Ingredients

1 pack organic tempeh

¼ cup fresh orange juice

1½ tablespoons rice wine vinegar

1 tablespoon maple syrup

2 tablespoons tamari

1 teaspoon grated ginger

1 clove garlic, minced

2 tablespoons extra virgin olive oil

1 lime

Handful of fresh coriander

Directions

1 If frozen, thaw 1 block of tempeh just enough to cut through, and cut lengthwise into 10 strips. Slice your tempeh into thin, bite-size pieces. Set aside in a small bowl.

2 In a separate bowl or a small Mason jar, mix together the orange juice, rice wine vinegar, maple syrup, and tamari. Grate the ginger and finely mince the garlic, and add to the mix. Shake and leave to sit. This is the base of your marinade, so be sure to mix well.

3 If you feel like frying:

Place olive oil in a large frying pan on medium to high heat. Add the freshly sliced tempeh and get ready for some sizzling as you watch your tempeh turn golden brown. Be careful to watch it closely and turn the tempeh so each side is cooked for approximately 5 minutes each. Add the orange-maple glaze from your bowl or jar to the hot frying pan. Turn down the heat to low-medium and watch the sauce simmer, get thicker and stickier, and turn into the lovely glaze you're aiming for.

If you feel like baking:

Place your tempeh into a casserole dish and pour the marinade over it. Marinate in the fridge for 4 to 12 hours. Bake in the oven at 350 degrees, covered, for 20 minutes, and then continue baking uncovered for approximately 10 minutes, or until the marinade is absorbed.

4 Whether you fry or bake your tempeh, the final touch is to add a little squeezed lime and coriander. The amount is up to you and your personal taste, but we suggest at least one-half of a small lime and a handful of fresh coriander on top!

Country Miso Soup

Prep time: 10 min • **Cook time:** 20 min • **Yield:** 10 servings

Ingredients	Directions
6 cups water	**1** In a large pot, bring 6 cups of water to a boil and then lower the temperature.
4 dried shiitake mushrooms, sliced	
½ ounce wakame (sea vegetable)	**2** Add the mushrooms, wakame, radish, celery, carrots, and onion.
1 piece daikon radish, cut into thick slices	**3** Cover and bring to a simmer, stirring occasionally for 20 minutes, or until the vegetables are tender.
2 stalks celery, cut into ¼-inch slices	
2 large carrots, peeled and halved lengthwise and cut into thick slices	**4** Take about 1 cup of liquid out of the pot and put it into a separate bowl. Add the miso paste and stir to fully blend in the miso.
1 small onion, cut into slices	**5** After the miso fully dissolves, add the liquid back into the pot with the heat turned off. Don't allow the soup to boil after you add the miso because doing so destroys some of the miso's nutrients.
1¼ cups dark brown or yellow miso paste	
1 head of bok choy, chopped	
3 green onions	**6** Stir in the bok choy and garnish with green onions.

Homemade Hummus

Prep time: 8 hr • **Cook time:** 2 hr • **Ferment time:** 7–12 hr • **Yield:** 2½–3 cups

Ingredients	Directions
2 cups chickpeas, soaked and cooked	*1* To prepare the chickpeas, rinse and soak in water for 8 hours. Rinse and drain and then place the chickpeas into a pot and cook with a piece of kombu for 2 hours on medium heat.
¼ cup tahini	
¼ cup lemon juice	*2* Once the chickpeas are finished, drain them through a colander and reserve some of the cooking water in a separate bowl.
2 cloves garlic, minced	
1 teaspoon ground, toasted cumin	*3* Combine the cooked and drained chickpeas, tahini, lemon juice, garlic, cumin, salt, and olive oil in a food processor and blend until smooth.
¾ to 1 teaspoon sea salt	
¼ cup olive oil	*4* With the motor running, slowly add the chickpea cooking liquid until you reach the desired consistency.
¼ to ½ cup reserved chickpea cooking liquid	
¼ cup vegan yogurt starter with water	*5* Put the hummus in an airtight container with the starter and leave sealed for 7 to 12 hours.
	6 Serve with sprouted spelt flatbread or veggie sticks.

Black Bean Salad with Fresh Mint

Prep time: 20 min, plus standing time • **Cook time:** 90 min • **Ferment time:** 2–3 days •
Yield: 1–2 quarts

Ingredients	Directions
1 cup black beans, soaked	*1* Drain the beans and rinse well with water. Drain again.
3 cups filtered water (no water needed if you use canned beans)	*2* Combine the beans and water and bring to a boil over high heat.
3 small red radishes, diced	*3* Lower the heat and cook for 1 to 1½ hours. Drain again.
3 scallions, minced	
¼ cup parsley, minced	*4* Place the beans, radishes, scallions, parsley, and mint in a glass jar and cover with water and salt.
¼ cup fresh mint, minced	
2 cups water	*5* Allow the beans to sit for 2 to 3 days.
½ to 1 tablespoon sea salt	
1 tablespoon mustard	*6* Whisk together the remaining ingredients and pour over the beans when you serve them.
1 tablespoon honey	
2 tablespoons lemon juice	
3 tablespoons apple cider vinegar	
½ to 1 teaspoon sea salt	
¼ tablespoon extra virgin olive oil	

Vary It! Instead of black beans, you can use kidney beans, black-eyed peas, garbanzo beans, or white beans. You can even mix a few different beans for a colorful and festive salad.

Sesame Miso-Crusted Tempeh

Prep time: 15 min • **Cook time:** 30 min • **Yield:** 8 servings

Ingredients	Directions
1 package organic tempeh	*1* Simmer the tempeh in a shallow pan with the water, tamari, and sliced ginger for 15 to 20 minutes. Remove and pat dry with a towel.
½ **cup water**	
1 tablespoon tamari	
A few slices of ginger (for simmering in pan)	*2* Place 1 tablespoon sesame oil in a glass dish and coat the bottom.
Marinade:	*3* In a small bowl, mix miso paste, honey, garlic, and mirin. Coat the tempeh completely with the mixture.
1 tablespoon sesame oil	
1 tablespoon miso paste	
2 tablespoons honey	*4* After you coat the tempeh with the marinade, dip the tempeh into sesame seeds and coat both sides.
2 cloves garlic, minced	
1 tablespoon mirin	*5* Cook at 350 degrees for 10 to 15 minutes.
½ **cup sesame seeds (black and white)**	

Note: Miso and tempeh are fermented, loaded with enzymes and protein, and digest really well. Together they make the perfect savory and hearty combination for a "meaty" addition to your meal.

Chapter 10

Nuts, Seeds, Coconuts, and Tubers

....

In This Chapter

▶ Learning all about nuts and seeds
▶ Understanding the benefits of soaking and sprouting
▶ Creating creamy dairy-free ferments
▶ Playing with potatoes and other tubers

....

*N*uts have a long history of being one of nature's richest sources of nutrients. Most are filled with protein, fiber, and heart-healthy fats while providing the body an essential source of energy. There are a variety of nuts; some grow on trees, some grow in the ground, and each has its own specific uses and benefits. What they all share in common is that they are delicious and lend themselves well to fermented recipes. These foods are essential as nondairy versions of long standing favorites such as creams, milks, cheeses, and yogurts. See what is in store for you below!

Nuts and Seeds: Great Nutrition in Small Packages

Nuts and seeds are bearers of life literally from the ground up. Whether they grow into an incredible fruit-bearing plant, a luscious vegetable, or are harvested in their whole form, they provide the basis of a wholesome, natural, and plant-based diet.

Selecting seeds

Seeds offer numerous benefits. They are allergy-safe, low in fat, high in protein, and are extremely versatile when it comes to everyday usage. Most people can also digest seeds much easier than nuts since they are smaller and the body seems to handle them very well.

- ✔ **Pumpkin seeds:** These small flat seeds have a slight crunch, interesting green color, and nutty flavor. Pumpkin seeds are higher in protein than other seeds as well as a high source of omega 3, iron, zinc, and phosphorous. They also contain calcium and vitamin B.

- ✔ **Sunflower seeds:** These seeds are found within the shell of the sunflower plant and are a whitish, pale gray in color. Once shelled, they can be substituted for nuts in many recipes. Sunflower seeds are a great source of energy and unsaturated fat. They contain calcium, protein, iron, and several vitamins.

- ✔ **Sesame seeds:** With a nutty taste and a delicate, almost invisible crunch, these seeds come in a host of different colors, including white, yellow, black, and red. Sesame seeds are highly active antioxidants and help to protect the liver from oxidative damage. They can act to reduce cholesterol and are also one of the best sources of calcium among plant-based foods.

- ✔ **Hempseeds:** These tiny, flat, yellowish green seeds have a soft texture and nutty taste. Hempseeds are a complete protein source as well as extremely high in essential fatty acids and calcium. They contain more than 20 trace minerals and are a great source of fiber and vitamin E while also containing chlorophyll. Store hempseeds in a sealed jar or a dark bag in the refrigerator.

Getting nutty

My top choices for fermented nuts are almonds and cashews. Both have a rich creamy texture and do well with soaking. They take on an exceptional taste and provide a robust, dense, and cheesy or creamy texture to any vegan recipe. Once soaked and fermented these nuts can be used to make cheeses, creams, dips, or simply used as a milk.

- ✔ **Almonds:** High in protein, fiber, calcium, and iron, almonds are also the richest source of vitamin E. They are also high in unsaturated fats. Almonds are fabulous for heart health and keeping your skin strong and supple. They are easiest to digest when they are soaked overnight. Store almonds in a glass jar in the pantry, but once they are fermented into a cheese, yogurt, or dip, be sure to seal in a glass jar in the fridge.

✔ **Cashews:** Cashews are a fatty nut, which is why we all love them so much! They are warming, sweet, and delicious. They contain protein, magnesium, phosphorus, and potassium. Cashews take well to soaking to make them easier to digest. They make fabulous tasting nut milk, creams, and spread. As an alternative to dairy cheese, cashews provide the texture and mouth feel that is often craved. Store them raw in a glass jar.

Some nuts can't be sprouted. Avoid walnuts, pecans, or any nuts that have been irradiated, roasted, or blanched.

Sprouting Nuts and Seeds

Nuts and seeds contain all the energy and nutrients needed to grow a plant in a concentrated form. It's as though they are asleep, waiting for the right conditions to allow them to grow. Sprouting a seed or nut wakes them up, and they begin to release their nutrients and activate their enzymes.

To make delicious sprouts, start with the best quality nuts and seeds you can find. Make sure they are raw, fresh, and organic, if possible. Watch out for nuts or seeds that have been roasted or irradiated — they won't sprout! And avoid any that are discolored, dried, or cracked.

When a nut or seed is soaked, some impressive things start to happen!

✔ **Increasing nutrients:** Sprouting increases B vitamins (especially B2, B5, and B6) and carotene, and produces vitamin C.

✔ **Fighting phytic acid:** Nuts and seeds contain something called phytic acid, which blocks the body's ability to absorb certain minerals, like calcium, magnesium iron, copper, and zinc. But the good news is that sprouting neutralizes phytic acid!

✔ **Neutralizing enzyme inhibitors:** All nuts and seeds contain enzyme inhibitors that keep them in a resting state until they have the right conditions to grow. Unfortunately, enzyme inhibitors interfere with our body's own digestive enzymes. Sprouting to the rescue! Soaking and sprouting nuts and seeds neutralizes enzyme inhibitors, making these little nuggets of nutritional goodness easier for us to digest. Sprouting also produces enzymes, giving our digestion a little helping hand.

Don't waste your time sprouting walnuts or pecans which have been removed from their shells and therefore will not sprout! Macadamia nuts are just difficult to sprout and may take up to 30 days!

Learning how to sprout

Sprouting couldn't be easier. The basic technique is the same for every sproutable nut and seed. The only difference is that some nuts and seeds take longer to sprout, so you'll have to be more patient. Here's what you need to do:

1. **Wash, then soak the nuts or seeds in water.**

   Soaking times vary so refer to Table 10-1.

2. **Drain, rinse, and put them in a jar.**

3. **Cover the jar with something that allows for air flow, but will also keep any dirt or bugs out.**

   A piece of cheesecloth or paper towel secured with a rubber band works nicely.

4. **Sprinkle with water twice a day until they begin to sprout.**

   You'll know it's working when you see little tails start to grow.

Instead of sprouting in a jar, you can buy a mesh bag designed specifically for sprouting or making nut milk. Or, try sprouting your nuts or seeds in an ordinary kitchen sieve set over a bowl.

Keep the nuts or seeds damp but not wet. You don't want mold to grow.

Table 10-1		Sprouting Times	
Nut or Seed	*Soak*	*Rinse*	*Sprout*
Almond	8–12 hours	Three times a day	Three days
Pumpkin	8 hours	Three times a day	Three days
Sunflower	2 hours	Twice a day	12 to 18 hours
Sesame	8 hours	Four times a day	Two to three days

Chart adapted from `http://wholerawfoodnow.com/articles/item/18-sprouting-chart` *and Nourishing Traditions (Fallon)*

Drying or storing your sprouts

Once soaked or sprouted, nuts and seeds should be stored in the fridge if they are not going to be used right away in a recipe. They will store for up to one week before they start to lose nutrition or grow mold in the fridge. Otherwise they can be placed in the freezer. You can also dry them in a warm oven or dehydrator, set at about 120 degrees, and test them until they are dry.

Using sprouted nuts and seeds

Sprouted nuts and seeds can be used in many different ways to "liven" up your meals (literally!). Here are some ideas:

- Add them to salads or sandwiches
- Use them in a homemade granola or muesli
- Make a super-powered trail mix
- Top off your stir-fry, curry, or other vegetable dish
- Make a flavorful raw nut and seed pâté or filling for a wrap (sprouted, of course)
- Get creative with nondairy nut cheese
- Blend or grind them to make a creamy nut butter
- Snack on them

Making Nondairy Ferments

You can make flavorful creamy, nondairy milks using a variety of nuts and seeds. To create a nondairy "milk," blend your choice of nuts or seeds with water in a ratio of 2:1 and strain out the pulp. These nuts and seeds are popular choices:

- Almonds
- Hempseeds
- Cashews
- Macadamias
- Brazil nuts
- Pumpkin seeds

Each will lend a different flavor and creaminess to your milk. Try combining more than one nut or seed together for even more variation.

Alternatives to yogurt makers

If you don't want to invest in a yogurt maker, try one of these alternatives:

- ✔ Wrap the jar in a dishtowel and put it in an insulated cooler with the cover on.

- ✔ Turn the pilot light on in an electric oven, then wrap the yogurt in a dishtowel and

put it inside the oven. (Do not turn your oven on!)

- ✔ Set the yogurt inside a food dehydrator that has the trays taken out and set for no higher than 110 degrees.

Making yogurt without the moo

Once you have your milk, add your starter and ferment as you would a more conventional dairy yogurt. You can use a dollop of store-bought yogurt, some sauerkraut juice, water kefir, or other "live" culture.

Make your own fruit yogurt blends by adding in some homemade fruit preserve such as the apple spice or peach cinnamon chutney. Or simply stir in some fresh or frozen berries and let them bleed into the yogurt for some natural color. Enjoy with some homemade granola.

Culturing nondairy cheese

Creating a nondairy cheese is similar to making yogurt. Use less water when making your base. The result will be a thicker, creamier product. Add culture, ferment for a few days, and you'll have a tangy cheese to spread on crackers, top your salads, or just enjoy with a spoon! For a firmer cheese, you can strain it through cheesecloth or even dehydrate it for a few hours to achieve a rind.

Add herbs, spices, or dried fruit to your "cheese" and pair them with sourdough bread, fruit chutney, or tangy pickled vegetables for a truly fermented feast!

Cracking into Coconuts

Coconuts are one of our favorite tropical foods to consume — whether in its whole form or as a milk, oil, or butter. All are delicious and good for you.

Coconuts are good for you!

Many people are still afraid of this saturated plant-based fat, but fear not. The fat found in coconuts is used by the body for energy, not stored as body fat. Coconut is now thought to be a source of exceptionally healthy oil for vegetarians and vegans. So enjoy it in many different ways — especially as a fermented yogurt!

Whole coconuts can be cracked open and used for the water, a rich source of electrolytes to drink, and the meat can be scooped right out. The meat is filled with healthy fat and fiber.

Be sure to choose young coconuts, as the meat will be soft. Mature coconuts have a hard, dried meat that does not taste as sweet.

When the water and meat are blended together, you get coconut milk. This provides the base for an excellent fermented nondairy yogurt. Coconut milk can be purchased in a can or in a carton. It is best to buy it unsweetened so that you can flavor your yogurt on your own.

If buying canned coconut milk, watch out for additives, sugars, or other stabilizers.

Coconut oil comes from the fatty acids pressed from coconut meat. Considered a "good" saturated fat, it is made up of medium chain fats that convert to energy. It is also stable oil, which means that it can be heated above 240 degrees.

Fermenting Potatoes and Other Roots

Both white and sweet potatoes can be fermented. We tend to lean towards the latter, as sweet potatoes are a rich source of vitamin A, beta carotene, and fiber, and just taste delicious. However, if they are not your thing, you can even get your hands on "white sweet potatoes," which look like a potato but taste like a sweet potato. Either way, all can be fermented!

They can be fermented raw, or even after they have been lightly cooked. The best way to ferment them is to use a basic brine, sauerkraut juice, or even coconut yogurt starter for extra flavor. You can do anything with them after, as a cooked potato will always taste better steamed, roasted, baked, or grated.

Preparing cassava

More commonly known to us as tapioca, cassava contains cyanogenic glucosides, which can form a toxic substance once ingested. So it must be prepared before eating. Some common techniques are to peel, cook, or grate it. One of the easiest ways is to ferment it in water for a few days.

Talking about taro

Taro is a root vegetable from the same family as cassava. Like cassava, it contains toxins and so must be cooked or fermented to make it an edible root. Once prepared, you can treat taro in a very similar fashion to potatoes. It is worth the effort for its delicious chewy, starchy, and versatile taste.

Taro root can irritate the skin. Be sure to wear gloves when peeling these tasty tubers!

Sunflower Seed Sour Cream

Prep time: 10 min • **Ferment time:** 1–3 days • **Yield:** ½ cup

Ingredients	Directions
1 cup raw sunflower seeds	**1** Soak sunflower seeds and hempseeds for about 8 hours in enough water to cover them.
1 tablespoon hempseeds	
4 tablespoons left over grain (brown rice)	**2** Drain off and reserve excess water, then puree soaked seeds with other ingredients in a blender or food processor. Add reserved water, just a little at a time, until the mixture reaches a thick, creamy consistency.
3 tablespoons olive oil	
1 tablespoon finely chopped scallions	
¼ teaspoon celery seed	**3** Place the mixture in a jar or nonmetal glass bowl and allow it to ferment for 1–3 days.
⅓ cup lemon juice	
1 tablespoon kefir grains	**4** Remove the kefir grains. If you are not able to find them, not to worry, they are completely edible.
	5 Enjoy sunflower cream on sweet potatoes, as a dip, or topped on some cultured black beans.

Cashew Sweet Cream

Prep time: 10 min • **Ferment time:** 1–3 days • **Yield:** 2 cups

Ingredients	*Directions*
1½ cups raw cashews	*1* Place the cashews and water in a bowl and let it sit overnight. In the morning, strain the cashews reserving the liquid.
1½ cups water	
2 tablespoons raw honey or coconut nectar	*2* Place the cashews in a food processor with the rest of the ingredients and blend on high for a few minutes until light and creamy.
1 teaspoon vanilla	
1 tablespoon lemon juice	*3* Place in a jar or small glass bowl and allow it to ferment for 1 to 3 days.
1 tablespoon kefir grains	
1 tablespoon nut butter (almond or cashew)	*4* Add some of the reserved soaking water or fresh water to the food processor and blend again to reach a desired consistency.

Cultured Nut or Seed Milk

Prep time: 10 min • **Ferment time:** 2 days • **Yield:** 4 cups

Ingredients	*Directions*
1 cup almonds, brazil nuts, or cashews, or for a nut-free variety try hempseeds, pumpkin seeds, or sunflower seeds	*1* Place the nuts or seeds in your blender along with water and blend on high speed for at least 1 minute.
4 cups water	*2* Strain the liquid through a cheesecloth, nut milk bag, or colander, pressing to squeeze the moisture from the nuts or seeds over a large bowl.
1 tablespoon coconut oil	
2 tablespoons raw honey, maple syrup, or coconut nectar	*3* Set the "pulp" aside and pour the liquid back into the blender along with the remaining ingredients except for the kefir grains.
1 teaspoon vanilla bean powder	
1 teaspoon soy or sunflower lecithin (acts as a binder)	*4* Blend until the liquid is emulsified.
1 teaspoon cinnamon	*5* Pour into a glass jar with the kefir grains and allow the "milk" to ferment for 2 days.
1 tablespoon kefir grains	

Tip: You can use your leftover pulp in a variety of recipes. Try subbing it for a cup of flour in a muffin, bread, or cracker recipe. They can even be used in raw desserts or sprinkled over your breakfast porridge. Extra fiber!

Note: If you are using any of the nuts, presoak overnight or for at least 4–8 hours.

Herbed Cashew Cheese

Prep time: 5 min • **Ferment time:** 1–2 days • **Yield:** 1½ cups

Ingredients	*Directions*
½ cups cashews soaked for at least 1 hour	**1** Drain the cashews and place in a blender with rejuvelac. Process until the mixture is smooth and creamy. You may have to scrape down the sides a few times.
2 tablespoons rejuvelac	
1 teaspoon dried basil	**2** Add more rejuvelac in ½ teaspoon amounts to get a desired texture. The puree should be thick and chunky with some texture.
1 teaspoon dried thyme	
1 teaspoon dried oregano	**3** Remove from blender and place in a cheesecloth and set in a colander inside a bowl. Cover the cheesecloth with another cheesecloth and allow it to sit for 1 to 2 days on the counter.
¼ teaspoon sea salt	
2 tablespoons olive oil	
	4 Be sure to taste it after 24 hours and allow it to ferment longer if you want a tangier taste.
	5 When it's at your desired flavor, combine the herbs, salt, and olive oil.
	6 Combine everything with a fork. Mold it together to form a "brick" or "log," or crumble it for more of a ricotta or feta texture. Refrigerate.

Almond Basil Pesto

Prep time: 5 min • **Ferment time:** 1–3 days • **Yield:** 2 cups

Ingredients	Directions
1 bunch fresh torn basil	**1** Place all ingredients except for the rejuvelac in a food processor and blend until smooth.
¼ cup chopped parsley	
1–2 cups whole almonds soaked overnight or for 8 hours	**2** Add in the two tablespoons of rejuvelac, mix thoroughly together, and allow the pesto to ferment for 1 to 3 days in the fridge.
½ cup pine nuts	
2–4 tablespoons lemon juice	
1 clove garlic	
¼ cup olive oil	
2 tablespoons rejuvelac	

Coconut Yogurt

Prep time: 20 min • **Ferment time:** 1–3 days • **Yield:** 4 cups

Ingredients	*Directions*
3–4 cups coconut milk (approx. 2 cans)	*1* Heat the coconut milk to approximately 115 degrees. As the milk cools to 110 degrees mix in the thickening agent of your choice.
Thickening agent (either 3 tablespoons tapioca or arrowroot, or 1–2 teaspoons agar flakes)	*2* Whisk in the thickening agent into a small amount of milk in a small bowl and then pour that mixture into the larger amount. Mix well to combine.
1 packet of nondairy yogurt starter	*3* Once the milk has reached 110 degrees, add the yogurt starter and mix well to combine it in fully.
	4 Incubate the mixture at 108 degrees to 112 degrees for 8 to 24 hours.
	5 Once the yogurt has solidified, allow it to cool for an hour or two until it reaches room temperature.
	6 Place the yogurt in the fridge for 6 or more hours to stop the culturing process. The yogurt will thicken as it chills.

Note: Coconut milk takes longer than animal-based milk. If you desire more of a sour flavor, culture it for a little longer.

Fermented Sweet Potatoes

Prep time: 20 min • **Cook time:** 45 min • **Ferment time:** 1–2 days • **Yield:** 2 cups

Ingredients	Directions
1 kilogram sweet potatoes (approximately 2 large or 4 small)	*1* Wash and peel your potatoes, cut into large chunks, and dry bake for about 45 minutes.
1 tablespoon sea salt 4 tablespoons coconut yogurt	*2* After they are cooked, place into a bowl with the sea salt and blend well. Stab with stab mixer. Add in the coconut yogurt.
	3 Place the mixture into a large glass jar with lid with room for it to expand. Leave in warm place for 1 to 2 days and then place in the refrigerator.

Tip: Sweet potatoes are vitamin A rich and full of antioxidants. They are naturally sweet and taste delicious even just plain. So why not make them last longer by fermenting them? This recipe tastes amazing on its own or as a side dish with cooked quinoa or brown rice alongside some steamed greens. Or try it for breakfast with more yogurt, some cinnamon, and honey!

Part IV
Meat, Dairy, and Eggs

Finding quality meat and dairy products can seem like a huge chore. Visit
www.dummies.com/extras/fermenting for tips on finding sources
of healthy, local foods.

In this part . . .

- ✔ Find out about the history of fermenting milks from around the globe.

- ✔ Gain an understanding of lactose intolerance and discover the health benefits of consuming fermented dairy.

- ✔ Explore the different culturing techniques used to create a wide range of healthy foods.

Chapter 11

Got Milk?

Despite the strange name, fermented dairy, also known as *cultured dairy,* is very common around the world. In fact, you probably have some in your refrigerator right now in the form of yogurt or sour cream.

Fermenting dairy provides not only delicious food but also improved nutritional value. Many people who have assorted digestive issues see improvement with regular consumption of fermented dairy.

The history of fermented dairy is long and a bit uncertain. There's a general consensus that it began thousands of years ago, when a nomad's pouch of milk was warmed just right on his camel, so that by the end of the day, it had become something delicious.

No matter the truth of the story, it would make sense. Placing a dairy product into a vessel (modern-day fermenting fans recommend a glass jar instead of a stomach pouch) and allowing it to sit in a warm place is the basic premise behind all fermented dairy recipes.

Fermenting milk not only makes it taste good but also helps preserve this delicate food. In many cultures, fermented dairy is the *only* dairy. You may find that this is true for your diet when you taste the results. In this chapter, we tell you everything you need to know to begin fermenting milk.

The Basics of Fermenting Milk

When you think about milk, the first type that probably comes to mind is cow's milk. Although in the United States cow's milk may be the most popular type of milk, fermenting has been around for thousands of years, and in the old days cows weren't as popular as they are today.

Which means that if you're just starting out fermenting your dairy, you can ferment milk from a variety of animals besides cows: goats, sheep, buffalo, and even more exotic animals, like camels and reindeer! Some fermented foods specify a particular animal to use, but this is usually so that the finished product has a certain flavor and consistency. Don't be afraid to experiment, especially if you have access to fresh milk from another type of animal versus store-bought cow's milk.

Making lactose tolerable by fermenting

You're probably familiar with people who are lactose intolerant. They become uncomfortably gassy or bloated, or even nauseous, if they consume dairy products. If you're unable to digest, or you have difficulty digesting, the natural sugar *lactose* found in milk and other dairy products, then you may be lactose intolerant.

Lactose intolerance happens because of a shortage of the naturally occurring enzyme *lactase* in your digestive system. This shortage is linked to certain genetics and can even be a result of the aging process. Lactase naturally breaks down lactose so that you can absorb it into your bloodstream.

Luckily, fermenting dairy can make it tolerable and even healthful to someone who's lactose intolerant. Through the fermentation process, the beneficial bacteria consume the lactose and turn it into lactic acid. There's also evidence that eating fermented foods on a regular basis helps reestablish a healthy level of beneficial bacteria in the gut, improving all digestion.

Many dairy products that are familiar on the store shelf can be naturally fermented and well tolerated by those who are lactose intolerant. Although this list is by no means exhaustive, some of the more common forms of fermented dairy include

✔ **Buttermilk:** True cultured buttermilk isn't a grocery store item. It's fermented with a variety of *lacto* bacteria that gives it the tang and health benefits of a fermented product.

- **Cheese:** The largest class of fermented milk products, cheese is created by using specific bacteria and molds developed for specific flavors and textures.

- **Cultured butter:** You create this product by adding some cultured buttermilk to the cream you churn. Allow it to ripen and then make your butter.

- **Kefir:** Kefir is created with bacterial grains that resemble cauliflower. These gelatinous grains are removed and reused again. They eventually grow in size and can be divided to share.

- **Koumiss:** This fermented, carbonated dairy drink was traditionally made from mares' milk in central Asia. Koumiss has been used medicinally in various times. Depending on the level of fermentation, it can contain varying degrees of alcohol. The recipe for the version in this chapter uses yeast to start the fermentation.

- **Sour cream:** Cultured sour cream is different from the product found on most grocery shelves. The fermenting process results in a deliciously acidic food, using *Lactococcus lactis,* a bacterium that's also used in the making of buttermilk and many cheeses.

- **Yogurt:** Yogurt is usually made from dairy milk and beneficial bacteria such as *Lactobacillus bulgaricus* and *Streptococcus thermophilus.* You can use a small scoop of finished yogurt to start a new batch, in an endless cycle.

Choosing pasteurized or raw milk

Many fermented food enthusiasts recommend using only raw milk products to make fermented dairy. Does this mean that any dairy you use must be raw? What is *raw,* anyway? Read on for an explanation of the different processes that milk undergoes before you purchase it.

- **Raw:** This means that the milk has gone through a filtering process and nothing else. No heat has been applied to the milk, and it will separate into a layer of cream and a layer of milk if left undisturbed. Raw milk still contains the enzymes that are naturally present, and instead of going sour over time like store milk, it simply clabbers, or ferments. Raw milk isn't legal to sell in all states, so check your local laws before buying it.

The argument against raw milk is that it may contain harmful bacteria and may transfer disease to the people who drink it. Although this is true in theory, raw milk farmers go through the same rigorous hygienic standards as farmers who pasteurize their products. In fact, some are held to even higher standards for bacterial counts than their pasteurizing neighbors. Advocates of raw milk believe that drinking raw milk with the natural enzymes and good bacteria still in it is healthier than drinking a sterilized milk product.

✔ **Pasteurized:** This is the process of heating milk to kill certain types of bad bacteria, such as *E. coli* or *listeria.* Milk is heated to 145 degrees for a period of time and then rapidly cooled.

✔ **Homogenized:** For this process, milk is agitated and filtered to break the naturally occurring fat into small enough particles that it no longer separates into a layer of cream and milk.

✔ **Ultra-pasteurized:** This type of milk is more common in Europe, although it has become more popular in the United States in recent years. In ultra-pasteurization, milk is heated to 280 degrees for 2 seconds and then rapidly cooled. Ultra-pasteurization makes the milk shelf-stable for up to 70 days from the day of processing, when it's packaged in sterile containers.

Not all milks are created equal. They're all nutritious, but the flavor and texture of the finished product can be wildly different. As you start fermenting your own dairy products, remember that they won't be the same as storebought examples. They won't contain thickeners and sweeteners like the products on the grocery shelves usually do.

If you find that the flavor of your fermented dairy is not to your liking, you can change the type of milk and see whether that helps, or you can ferment for a shorter period of time. The longer you leave a food to ferment, the tarter it becomes.

A lot of enthusiastic people enjoy making and eating fermented dairy products, so try looking online for forums that match what you're interested in. Many of them list suppliers and can connect you with someone who may even be local. You may find companies that sell or trade dairy cultures for a reasonable price.

Separating milk

Many people have grown up hearing that whole milk is too high in fat for a healthy diet. They've become used to drinking low-fat milk, believing that it gives them the health benefits of milk without the harm of too much fat.

However, many health advocates believe that the healthiest milk is full-fat *and* fermented, in moderation.

Some recipes require the separation of cream and milk, and if that's how a recipe is listed, following the recipe is best. You can separate some milk by allowing it to sit without movement for a few hours or overnight. Cow's milk is like this. The next morning, you have a thick layer of cream atop the lower-fat milk underneath. This lower-fat milk is still not the nonfat variety sold in stores, but it is lighter tasting.

However, you shouldn't leave goat milk to separate. The naturally occurring caprylic acid increases over the life of the milk, and you end up with a goaty-tasting product. (Sorry, but that's the best way to describe it.) Avoid this by using a cream separator turned to the lowest setting to remove the cream from goat's milk.

Even if you don't mind full-fat milks, many recipes use just the cream or just the separated milk as an ingredient. Always be on the lookout for a cream separator when you start out fermenting dairy. They can be expensive if purchased retail, but many deals crop up from time to time.

Sourcing Your Starter Cultures

After choosing your milk, your next step is to find out where to buy the specific cultures needed to make your own fermented dairy products.

If you're trying to start your home-fermented products by using store-bought varieties, check that the beneficial bacteria are present. You need a living food to use as a starter for the next batch of living food. If you're shopping at a health-food store, tell the clerk you're looking for live cultured products. These products may have to be processed according to your state's food laws.

Another option is to buy a fermented food culture from someone who's also fermenting. Ask for cultures that have been used in your specific type of recipe. For example, kefir grains are grown as water kefir, goat milk, and cow milk. The grains of the water kefir aren't interchangeable with milk kefir. The goat milk grains may not do as well in cow's milk, and vice versa.

If you do find that you only have goat milk grains available and want to use cow milk, just know that the grains may take a few cycles to really start to grow. They certainly won't fail, but they can take longer than expected to ferment in the new type of milk at first.

The nice thing about fermented foods is that, after you have a batch going, you then have the future starter to make more indefinitely. Saving a bit from one batch to the next also creates a specific culture that loves where you're keeping it!

Keep in mind that starters are living bacteria. They're best kept at the proper temperature and fed regularly.

Culture overload

Even the most enthusiastic fermented food lover will eventually have more starter than he or she can use.

I (Amy) was once fermenting for my family and a friend. Both families drank 2 gallons of fermented kombucha a week. Every week, my culture doubled in size, and because I had 4 gallon jars working at once, I was overrun with excess culture! It was a nice problem to have, but still, I gave cultures away to everyone who came to my home, whether the person wanted one or not.

Many fermented food enthusiasts enjoy trading starters for little more than postage. Find the most local supplier to avoid your culture losing strength because of age or lack of food. Take advantage of this goodwill gesture and pay it forward to someone else after you get your own batch of fermented dairy going.

Beware of sites posing as experts with superior starters. If you find the cost to be prohibitive, keep looking. And don't try to sell your dairy starters for a hefty profit, because in the world of fermented food enthusiasts, doing so is in bad taste, and you may soon find you have no customers.

Serving and Storing Fermented Dairy Products

Fermented dairy products are cultured at room temperature or above. After they reach the desired fermentation, you chill them to slow the fermentation. For fermented dairy recipes, follow the recipe to the letter, taste it, and decide whether to stick with the same storage time based on how well you like the flavor. Though properly fermented food doesn't go bad, it does eventually become too sour-tasting to enjoy eating.

Serve fermented dairy products at their optimal flavor, especially when you're introducing someone to the idea of home-fermented items. For someone who's new to the taste, the tanginess may be off-putting if the product is overripe.

When serving things like yogurt and kefir, many people expect a highly sweetened taste. To retrain their palate, add natural sweeteners like honey, maple syrup, or organic sugar to the recipe. You can slowly cut back as their taste develops for the unsweetened product.

Making the most of your results

The one thing to remember about culturing foods at home is that they're best used continuously. To get into the habit of doing this, these foods should become important in your diet. The following are some ways to use fermented dairy foods:

✔ Use kefir in smoothies. Try using natural sweeteners to entice your family's taste buds. Reduce the sweetness as they become used to the flavor.

✔ Culture your own cream into sour cream and use it in place of store-bought sour cream.

✔ Create fresh cheese at least once a week and use it for snacks and dips, as after-school snacks, or as the first course for dinner.

✔ Eat yogurt with fresh fruit or granola as a healthy breakfast.

✔ Place yogurt in a cheesecloth-lined strainer and allow liquid (whey) to drain away. The resulting cheese-like product makes wonderfully rich-tasting, cream cheese–style food.

When starting out with any new food, moderation is key. Try eating (or drinking) your new food in small amounts to gauge your reaction to it. Although you should never feel ill, fermented foods can produce gas or slight bloating in some people at first. A good rule is to start with no more than ¼ cup of fermented dairy product a day for the first week. Then increase the amount the following week to give your system time to adjust.

If your fermented dairy seems too strong or sour, try reducing the time of fermentation for the next batch. Chilling it, sweetening it, and adding it to other foods are also good ways to help you get used to the natural tang.

As with any fermenting, use nonreactive containers for the products in this chapter. We discourage plastic; glass is the perfect vessel for most of your fermenting. It's easy to clean, it doesn't harbor odors, and you can recycle it. Other choices are stainless-steel and enamel-coated containers or lead-free crocks.

Fermented Sour Cream (Crème Fraîche)

Prep time: 10 min • **Ferment time:** 12–24 hr • **Yield:** 1 pint

Ingredients	*Directions*
1 pint heavy whipping cream (not ultra-pasteurized) ¼ teaspoon mesophilic Aroma B culture, dissolved in ¼ cup water	**1** Pour the cream into a sanitized jar, place the jar in a pot, and add warm water to the pot up to the level of the cream.
	2 Heat the pot gently, raising the cream temperature to 86 degrees. Stir the cream with a long-handled spoon as you warm it.
	3 Add the Aroma B starter solution and stir gently and thoroughly.
	4 Check the water bath temperature and adjust it to 86 to 88 degrees (add ice cubes or hot water, as needed).
	5 Place the lid loosely on the jar. Place the jar and pot in a warm place. An empty picnic cooler provides good insulation and keeps them warm enough. Let stand for 12 to 24 hours.
	6 Gently stir your new crème fraîche; it will be about as thick as creamy yogurt. You can use it right away or chill it for a few hours for a thicker, spreadable consistency and tangier flavor. Keep it in the jar in the refrigerator for up to 2 weeks.

Tip: Find the Aroma B starter at any cheese-making supply website. I (Amy) have had luck with New England Cheesemaking Supply Company (www.cheesemaking.com).

Koumiss

Prep time: 30 min • **Cook time:** 30 min • **Ferment time:** 2 days • **Yield:** About 1 gallon

Ingredients	Directions
1 gallon goat milk **2 tablespoons honey** **⅛ teaspoon champagne yeast** **¼ cup warm water**	*1* Heat the goat milk to 180 degrees in a stainless-steel pot and remove any film that forms. Add the honey and allow the mixture to cool to 70 degrees.
	2 Dissolve the champagne yeast in 115 degree water and let it stand for 10 minutes. Stir the yeast mixture into the milk. Cover it with a clean cloth and allow it to stand at room temperature until it foams (about 24 hours).
	3 Pour the koumiss into sanitized beer bottles that can withstand carbonation pressure. Fill only to 1 inch below the bottom of the neck of the bottle.
	4 Store the koumiss at room temperature for 24 hours and then refrigerate. Shake the bottles gently every few days but not just before opening.

Tip: You can buy the champagne yeast at a beer maker's shop.

Tip: Koumiss keeps for 6 to 8 weeks but then becomes increasingly acidic. You can add honey to hide the acidic taste.

Cultured Buttermilk

Prep time: 5 min • **Ferment time:** 24 hr • **Yield:** 1 quart

Ingredients	Directions
1 cup cultured buttermilk 3 cups whole milk	**1** In a sterilized glass jar, combine the buttermilk and whole milk and mix well.
	2 Cover loosely and leave at room temperature for 24 hours.
	3 Taste for desired tanginess, and refrigerate your buttermilk for up to 2 weeks.

Cultured Butter

Prep time: 10 min • **Yield:** 2 ounces

Ingredients	*Directions*
2 cups crème fraîche (see the earlier recipe)	*1* Pour cold crème fraîche into a large mixing bowl.
⅛ teaspoon salt (optional)	*2* Using a cold whisk or a mixer with a paddle attachment, vigorously mix the crème fraîche.
	3 After about 10 minutes, the butter will form into a lump and separate out of the buttermilk.
	4 Remove the cultured butter from the buttermilk and place it into a container with a cup of cold water. Continue pressing and folding the butter gently to wash away as much buttermilk as you can. Doing so helps the butter last longer.
	5 Sprinkle the salt (if desired) over the rinsed butter and gently fold it in.
	6 Store the butter in the refrigerator for up to 5 days.

Homemade Yogurt

Prep time: 20 min • **Ferment time:** 7 hr • **Yield:** 1 quart

Ingredients	Directions
2 quarts cow or goat milk **2 tablespoons yogurt that contains active cultures**	**1** In a 6-quart pot, heat the milk to 185 degrees. Don't allow the milk to boil over. Cool the milk to 110 degrees by placing the entire pot into an ice-filled sink and swirling it around.
	2 Add the yogurt to the cooled milk, stirring well. Pour the yogurt into the container that you'll be storing it in and place it in a dehydrator with an accurate thermometer gauge. Keep the yogurt at 110 degrees for 7 hours, or until it thickens and is as tangy as you like.
	3 Refrigerate the finished yogurt until ready to eat. Store it for up to 2 weeks in the refrigerator. Keep 2 tablespoons of this yogurt as the starter for the next batch.

Note: Goat's milk yogurt is naturally looser than cow's milk yogurt. This is normal.

Kefir

Prep time: 20 min • **Ferment time:** 24 hr • **Yield:** 1 quart

Ingredients	Directions
4 tablespoons kefir grains **1 quart milk**	**1** Place the grains in the bottom of a quart canning jar. Pour the milk over the grains, leaving ½-inch head space.
	2 Cover the jar with a coffee filter and place an elastic band around the filter, if desired. Place the jar in an out-of-the-way place at room temperature.
	3 After 24 hours, the kefir is ready. Strain out grains for the next batch and refrigerate the finished kefir. Store the kefir in the refrigerator for up to 1 week.

Tip: You can find kefir grains by doing a search online. A huge community of kefir drinkers willingly shares grains through the mail.

Note: Kefir is carbonated at first. Be aware that it can bubble over if agitated too much after it's finished. Never use metal utensils in any part of kefir making.

Strawberry Banana Kefir Smoothie

Prep time: 5 min • **Yield:** 2 servings

Ingredients	Directions
2 cups kefir	**1** Pour the kefir into a blender. Peel the bananas and cut them in half. Add the bananas, frozen strawberries, and honey to the blender.
2 large bananas	
2 cups frozen strawberries	
1 tablespoon honey	**2** Cover and blend until smooth and thick. Kefir smoothies can keep for up to 1 week in the refrigerator.

Vary It! Try this recipe with a sprinkle of cinnamon.

Pumpkin Kefir Smoothie

Prep time: 5 min • **Yield:** 2 servings

Ingredients	Directions
2 cups kefir	**1** Pour the kefir into a blender. Add the pumpkin puree, pumpkin pie spice, vanilla, and ice.
¼ cup pumpkin puree	
⅛ teaspoon pumpkin pie spice	**2** Cover and blend until smooth and thick. Kefir smoothies can keep for up to 1 week in the refrigerator.
¼ teaspoon vanilla	
4 ice cubes	

Vary It! Add a medium banana for extra sweetness.

Tip: The thickness of kefir varies from batch to batch — sometimes it's thick; other times it's thin. The thicker the kefir, the thicker your smoothie will be.

Chapter 12

Making Cheese

Cheese is popular the world over, but most people never think about how it gets to store shelves. Making your own cheese is actually fermenting in a very controlled manner.

What makes cheese delicious is the relationship between the flavor of the milk itself and the bacteria introduced into it. This means that after you get a cheese technique down, you can change certain ingredients and end up with an entirely different flavor.

Purists agree that there's only one way to make real mozzarella, and that involves buffalo milk, of all things. However, we believe you can certainly make a wonderful version without the buffalo milk, as the recipe later in this chapter demonstrates.

In this chapter, we show you how to ferment cheeses and help you expand your tastes horizons using simple, easy-to-find ingredients.

Understanding Cheese Making Ingredients and Techniques

Making simple cheese isn't that difficult, but the language that's used to describe the parts of the process can be a little confusing. There's no one way to make all cheeses, and by understanding the process you'll know whether you have the time and technique to tackle each new recipe.

Choosing milk

Despite the dairy cow being king, er, queen, you can make cheese from virtually any milk. Depending on the country of origin, certain cheeses are historically made from a specific species of milk or even a certain species that's raised in a specific location. That's dedication.

For hobbyists interested in fermenting their own cheese, most modern-day recipes recommend the type of milk to use and also what alternate milks can be used.

The ingredient that's not always interchangeable is the added amount of cream. When a cheese recipe asks for both a measure of milk and cream, you must add both. This sometimes results in a crossing of species, such as goat's milk and cow's cream. Your cheese may not pass the cheese purist's review, but it will still taste delicious.

All about rennet

Rennet is the magic that makes cheese happen. *Rennet* is an enzyme in the stomach of ruminants that have yet to start grazing or chewing their own cud.

Rennet is sold in stores in liquid and tablet form, and your recipe will recommend one or the other. Many times, the rennet is interchangeable, but breaking off a fraction of a tablet, instead of measuring out the equivalent in drops of liquid, may be difficult.

When purchasing rennet, look for it in the refrigerated section of the store. It's usually in the dairy section, although some stores keep it in the meat department.

Turning milk to cheese

Despite how technical modern cheese making has become, it all comes down to a few basic steps:

1. **Bring the milk up to temperature (to simulate the animal's body temperature, which activates the starter culture) and add the starter culture.**

2. **Add a coagulant, such as rennet.**

3. **Form and mold the curd, and drain the whey.**

4. **Salt the cheese.**

5. **Age the cheese (during which rind develops, with the exception of fresh cheeses).**

Some recipes omit one or more of these steps, but all cheese requires the milk to at least be brought up to temperature so it can coagulate. This is where the proteins clump together to form curds.

Fermenting

Cheese making is fermenting, although cheese isn't often put in the same category as other fermented dairy products like kefir. The idea behind creating a cheese through fermenting is the same: You introduce the proper bacteria (or culture) to the milk, and eventually the proper fermenting happens, and you have a fairly predictable flavor.

Cheese is often easier to digest for those suffering from lactose intolerance. Some people can tolerate only certain types of cheese.

Because cheese is a living product, the flavor changes over time, and like other fermented foods, you may find that you prefer a specific age of cheese. Adding more or less time may make that same cheese not palatable to you.

Colorings

Cheese comes in many colors. To some people, a cheese just doesn't look right without the golden or downright orange glow that comes from the addition of cheese coloring.

To create a yellowish (or darker) tinge, you can add drops of natural *annatto* (a liquid food coloring found in most cheesemaking supply stores) per the instructions in the recipe. Be careful — a tiny bit goes a long way! Usually you need only a drop or two. As the cheese dries, the color deepens.

Too much coloring is not only unappealing but can also make your cheese taste bitter.

If a cheese recipe calls for you to add a coloring, it's for aesthetics. You can make your cheese without the coloring and it will still turn out fine.

Cheese can also have the natural coloring that comes from washes, like beer and wine, or the mold that grows on and inside of the cheese. These colors are also indications of a recipe's success and will be clearly stated in the directions.

Salting

Salting is the next step in most cheese production. You apply salt either directly onto the cheese exterior or as brine that you wash over the cheese.

Salt slows down enzymatic activity, enhances flavor, and keeps unwanted organisms away, such as bacteria that can ruin the cheese or sour the taste. Salt also draws out moisture and helps form the *rind,* or the outside of the cheese.

Ripening

Ripening, aging, and *fermenting* all refer to the period of time required for flavor to develop in your cheese. Depending on the type of cheese, ripening is a delicate play of time and taste.

This stage needs to be done in a controlled environment because, just like the regular rules of fermenting, you're trying to grow the bacteria you want and avoid giving the unwanted bacteria a chance. Every style of cheese requires a different treatment.

Storage

You need to know how to store your cheese both while it ripens and after it's ripe.

To store cheese while it ripens, cool is the rule. Keep your cheese storage area (called a *cheese cave*) cooler than your living space; 50 to 55 degrees is average. Using a root cellar is ideal, but some people use a dedicated refrigerator that they've converted to a cheese cave. You can find plans online to convert a refrigerator in this way through a simple web search.

After cheese is properly ripened, you need to store it properly as well. Here are some suggestions for storing all types of cheese:

- ✔ **Soft cheeses** have a short life span of a week or less. Most important is to prevent them from oxidizing, so keep them sealed in their original container or tightly wrapped in plastic wrap.

- ✔ **Semisoft, surface-ripened, semihard, and washed-rind cheese** should be wrapped loosely in parchment paper and placed in a plastic container with a tight-fitting lid. Store these cheeses in the vegetable crisper drawer of your refrigerator. Because cheese continues to ripen as it ages, release the air out of the cheese every day by unwrapping it and letting it sit at room temperature for 30 minutes.

✔ **Blue cheese** must be wrapped in wax or butcher paper and stored in a plastic container with a tight lid. You should also place it in the vegetable crisper drawer of the fridge.

Making Soft and Semisoft Cheeses

Soft and semisoft cheeses are wonderfully versatile, just right for snacking or cooking. The term *soft* refers to the amount of moisture left in the curds. Soft cheeses can be spreadable, gooey, or stretchy.

The main thing about soft and semisoft cheeses is that you must consume them pretty quickly. They're a bit more delicate than the drier, harder cheeses, and they don't have a long shelf life — the high moisture and low salt content can provide a breeding ground for bad bacteria and mold.

Soft and semisoft cheeses are often made within hours of milking. They rely on the freshness of the milk to shine through, and their creaminess is unmatched. Sometimes called *fresh cheese,* this category of cheese can be very simple to make for the hobby cheese maker and is a natural starting point in cheese making.

Making Hard Cheeses

After mastering soft cheeses, hard cheese is an easy jump. Most of the steps are the same with the exception of additional heating and drying time to develop the complex flavors and textures of hard cheese.

Just like soft cheeses, hard cheeses are commonly made from the milk of cows, sheep, or goats. After you figure out cheesemaking techniques, something as simple as changing the milk can bring new life to an old recipe.

Cheese starters for hard cheeses are usually rennet and beneficial bacteria, which helps develop the milk's flavor and continues to develop flavors as the cheese matures.

Cheese molds and bacteria are what add specific flavors. Here are some:

- **Penicillium roqueforti (mold):** Produces the blue veins in blue cheeses
- **Penicillium candidum (mold):** Produces the white mold on bloomy-rind cheeses like brie
- **Geotrichum candidum (mold):** Results in the wrinkly, mold-ripened cheeses
- **Brevibacterium linens (bacteria):** Produces the orange-red rind on stinky cheeses
- **Propionibacterium freudenreichii (bacteria):** Produces holes inside Swiss-style cheeses

Brevibacterium linens (or B. linens), which is used to make washed-rind cheeses, occurs naturally on human skin. That's why some stinky cheeses have a locker room smell to them.

Serving Cheese

When serving cheese, no rules are written in stone, but following certain guidelines will result in the best-tasting cheese experience. Cheese is easy to serve, and there's an assortment of tastes for every palate. Combine your selection with fruits and nuts that complement the cheese, and your guests will appreciate your effort.

Because milk is seasonal, its taste and components change throughout the year. During the early season, milk is richest because of the abundance of grazing for the animals. It's also highest in butterfat, solids, and protein. Mid-season milk has a higher water content and is lower in the other components. Finally, late-season milk is rich but lowest in protein and best for making hard cheese.

I (Amy) make cheese for my family, and I notice a clear difference in the results throughout the seasons. Early in the spring, I don't require as much citric acid to make my mozzarella stretch, and during the year, I make notes on how much citric acid is needed for which month.

When buying cheeses for tasting, take note of the cheese variety and season in which it's produced. You'll find that the same cheese made in a different locale or during a different time of year has a distinctly different flavor.

Fresh cheeses have a short shelf life, so you must eat them as soon as possible. Prepare your platter with all the additional things served with the soft cheese, and then add the cheese and serve immediately.

Hard cheese needs to come to room temperature, and you must cut it into manageable pieces. Even if the piece of hard cheese is to be cut on the platter, the wedges should be small enough to not touch each other, and give room for the cutting.

Most stores that have a wide selection of cheese also have someone who knows more than the average person about tastes and pairings. Ask the shopkeeper for suggestions, and both your guests and you will have a rich, new tasting experience.

Cottage Cheese

Prep time: 30 min, plus standing and storing • **Cook time:** 45 min • **Ferment time:** 15–20 hr • **Yield:** About 4 cups

Ingredients	Directions
1 gallon pasteurized whole cow's milk	*1* In a double boiler, slowly heat the milk to 72 degrees.
½ teaspoon mesophilic direct-set starter culture	*2* Remove from heat and stir in the mesophilic starter culture.
4 tablespoons heavy cream	*3* Cover and place in an out-of-the-way place, wrapped in a towel to help keep the pot at 70 to 72 degrees.
¼ teaspoon fine salt	*4* Allow the curds to sit for 15 to 20 hours. They'll become firm during this time.
	5 With a long, stainless-steel knife, cut the curds into ⅜- to ½-inch cubes and allow them to settle for 30 minutes.
	6 In a double boiler, slowly heat the curds over 45 minutes, stirring every 5 minutes to prevent them from sticking together. Your goal is to raise the temperature to 110 degrees during this time.
	7 Keep the curds at 110 degrees, and gently stir for 25 to 35 minutes more. The curds will start to firm up during this time.
	8 Remove the pot from the heat and allow the curds to settle for 10 minutes.

9 Line a colander with cheesecloth and place it in a clean pot to drain for 10 minutes.

10 Tie the ends of the cheesecloth together and dip the curds into clean water three times to rinse and firm them.

11 Drain the cottage cheese again in the colander for 15 minutes more.

12 Place the drained curds into a clean bowl and gently stir in the cream and salt.

13 Refrigerate for 2 hours before serving.

Cream Cheese

Prep time: 10 min • **Ferment time:** 36–48 hr • **Yield:** 1 cup

Ingredients	*Directions*
3 cups whole cow milk	**1** In a double boiler, combine the milk and cream. Gently warm to 72 degrees.
3 cups whipping cream	
½ teaspoon mesophilic starter	**2** Add the starter and stir gently. Add the rennet and stir gently once again.
2 drops liquid rennet	
	3 Cover the pot and place it in an out-of-the-way location where it won't be disturbed.
	4 Allow the mixture to ripen for 24 to 36 hours, or until milk has thickened to a yogurt consistency.
	5 Line a colander with cheesecloth and pour the curds into the colander. Bring up the corners of the cheesecloth and tie them together to create a bag.
	6 Hang the bag of curds for 12 hours more, or until the whey stops dripping off.
	7 Refrigerate the curds for 1 hour before serving to allow flavor to develop.

Note: Mesophilic starter is in powder form, and found at any cheesemaking supply stores.

Fermented Garden Vegetables, Brined Asparagus, and Garlic Dilly Beans

Lacto Fermented Eggs and Preserved Lemons

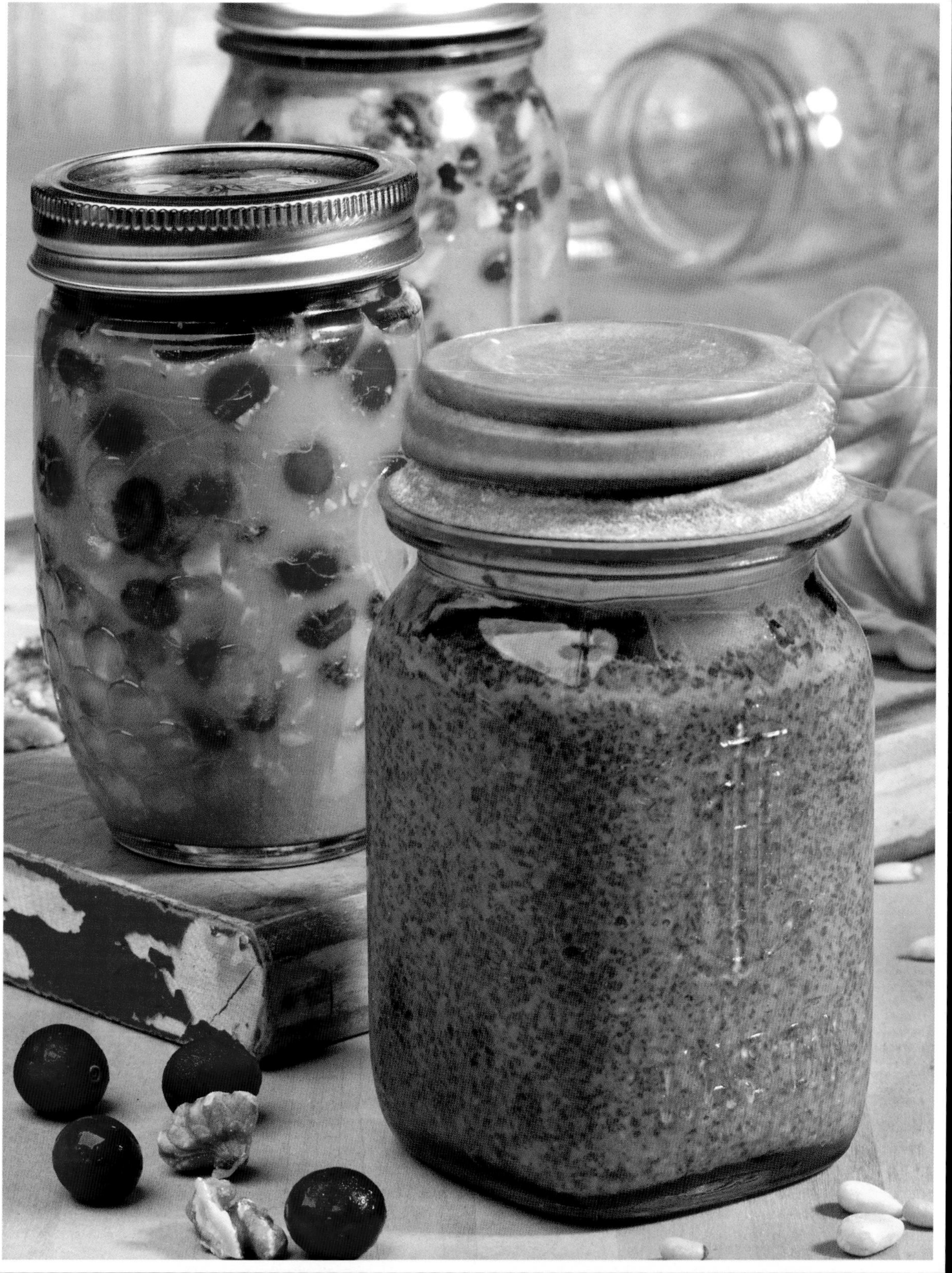

PHOTO COURTESY OF T.J. HINE PHOTOGRAPHY

Chevre

Cook time: 10 min • **Ferment time:** 60 hr • **Yield:** 4 small rounds

Ingredients	Directions
½ gallon pasteurized or raw goat milk (not ultra-pasteurized) ⅛ teaspoon MA4001 or similar freeze-dried, direct-set mesophilic culture ¼ teaspoon liquid calcium chloride mixed with 1 tablespoon water ⅛ teaspoon liquid rennet mixed with 1 tablespoon water Flaked or kosher salt	**1** Add the milk to a 3-quart, stainless-steel pot and slowly bring it to 86 degrees. Remove the warmed milk from the heat. **2** Sprinkle in the culture, wait 5 minutes, and then stir with 20 gentle strokes. **3** Add the calcium chloride-water mixture and stir gently. **4** Add the rennet mixture, stir, and cover. Let the mixture stand at about 72 degrees for at least 12 hours. Don't agitate or stir the mixture while the curd sets. **5** Place a draining rack inside the drain pan and place the forms onto the draining rack. Ladle the curd, which looks like thickened yogurt, into the forms. Allow the curd to drain for 24 hours at room temperature, pouring off the whey from the drain pan as needed to keep the cheeses above the liquid. **6** Remove the cheeses from the forms and place directly on the draining rack. Sprinkle on all sides with salt. Let the cheeses dry for another 24 hours at room temperature or refrigerate if it is especially warm weather, turning once or twice. **7** If desired, roll the cheeses in herbs before wrapping in wax paper, then wrap in plastic wrap. Refrigerate for up to several weeks.

Note: Direct set culture can be found at any cheesemaking supply store.

Queso Blanco (White Cheese)

Prep time: 20 min • **Drain time:** 2 hrs • **Yield:** 1 pound

Ingredients	Directions
1 gallon cow or goat milk ½ cup cider vinegar or lemon or lime juice	**1** Heat the milk in a 6-quart pot over medium heat. Stir it often to prevent scorching. When the milk becomes steamy and bubbles are just forming around the edges of the pan, turn off the heat.
	2 Pour the vinegar into the milk and stir gently until curds form. After the curds start to form, cover the pan and let it sit until you see a deep layer of whey on top with the mass of curds at the bottom of the pan.
	3 Place a linen- or cheesecloth-lined colander into the sink and scoop the curds into it. Discard the remaining liquid in the pot. Tie the corners of the cheesecloth together and hang until it stops dripping. (You can do this right in the sink or in the refrigerator if room is a factor.) Your cheese is ready to use!

Vary It! For a dessert cheese, use lemon or lime juice in place of the vinegar.

Ricotta

Prep time: 20 min • **Ferment time:** 12 hr • **Yield:** 1¼ cups

Ingredients	*Directions*
½ cup heavy cream 4 cups whole milk (not ultra-pasteurized) 2 tablespoons fresh-squeezed lemon juice Pinch kosher salt	*1* In a nonreactive saucepan, combine the cream, milk, and lemon juice. Cook over medium-low heat, stirring constantly to prevent scorching, until the mixture reaches 205 degrees.
	2 Remove the saucepan from the heat and let the mixture rest for about 15 minutes. During this time, the curds and whey will separate.
	3 Line a strainer with cheesecloth and set the strainer over a bowl. Ladle the curds into the strainer to drain the whey. Cover the strainer and bowl tightly with plastic wrap. Then refrigerate overnight to let the whey drain.
	4 Discard the whey and wipe the bowl dry. Transfer the ricotta to the bowl. Stir in the salt, cover the cheese tightly, and refrigerate until needed. Alternatively, you may transfer the ricotta to an airtight container. Refrigerated, this cheese will keep for up to 3 days.

Mozzarella

Prep time: 25 min • **Cook time:** 30–40 min • **Yield:** 2 pounds

Ingredients	*Directions*

Ingredients

2 gallons whole raw or pasteurized milk (not ultra-pasteurized)

1 tablespoon citric acid, dissolved in ¼ cup water

½ teaspoon liquid rennet, mixed in ¼ cup water

½ teaspoon lipase powder, dissolved in ¼ cup water, and set aside for 10 to 20 minutes

Kosher salt

Directions

1 Using a double boiler, warm the milk to 55 degrees. Gently stir in the dissolved citric acid. Then stir in the dissolved lipase.

2 Slowly heat the milk to 88 degrees over low to medium heat. The milk will begin to curdle.

3 Stir in the rennet as you raise the temperature of the milk to 100 to 105 degrees. Within a few minutes, you'll begin to see more curds forming. After you reach the appropriate temperature, turn off the heat and remove the pan from the stove. When the whey is relatively clear, the curds are ready.

4 Remove the curds from the whey by either pouring them out through a strainer or ladling the curds into a bowl. Whichever method you choose, reserve the whey in a separate stock pot.

5 Now you need to begin the kneading process that will make the mozzarella nice and elastic. To do so, heat the reserved whey to 170 to 180 degrees. Wearing gloves, shape the curd into a ball, set the ball on the ladle, and submerge the ladle into the whey, dipping it as needed (you want the curd to be smooth and elastic in texture). Knead the curd again and resubmerge into the hot whey. Continue kneading and redipping the curd until it reaches 145 degrees, the temperature at which the curd will stretch.

6 When the curd is the right temperature (145 degrees), it's ready to be stretched. Sprinkle a small amount of kosher salt on the cheese (as you would seasoning food to flavor it) and then fold the salt into the cheese. Continue to knead and pull the cheese until it's smooth and elastic. When the cheese pulls with the consistency of taffy, it's done.

7 Shape the mozzarella into small or large balls. You can eat it while it's still warm, or you can place the balls in ice water for about 5 minutes to quickly bring the inside temperature of the cheese down. Hot or cold, the cheese is best served fresh. For storage, place the balls in a freezer bag with a few tablespoons of milk and keep refrigerated. Eat within a few days.

Muenster

Prep time: 2–3 hr • **Ferment time:** 7 days • **Yield:** 1 pound

Ingredients	Directions
1 gallon whole goat's milk	**1** In a double boiler, warm the milk for 15 to 20 minutes, until it reaches 88 degrees. Remove from heat and allow to sit for 5 minutes.
¼ teaspoon rennet blended with ¼ cup cool water	**2** Add the rennet into the water and stir into the milk. Cover the pot and let set for 1 hour, or until the curd breaks cleanly when tested with a knife.
2 teaspoons fine sea salt	**3** Cut the curd into 1 inch cubes and sprinkle them with salt. Return the double boiler to the heat and work the salt into the curds with a long handled spoon. Don't stir; just turn the curds over gently with the spoon.
	4 Scoop the curds into a ball and place them in a colander lined with cheesecloth. Drain for 20 minutes.
	5 Ladle the curds into a 1-pound cheese press lined with cheesecloth. Apply 40 pounds of pressure for 12 hours. Remove the cheese, flip, rewrap, and return to the press at 40 pounds of pressure for another 12 hours.
	6 Rub the round of Muenster with a little salt and place on a mat for drying. Lightly salt and flip around once a day for 7 days.

Chapter 13

Meat, Fish, and Eggs

In This Chapter

▶ Discovering the types of fermented meats

▶ Choosing the best meats

▶ Making the decision whether to nitrate

Fermented meats started as a result of not having refrigeration methods to store meat products. Whether meat was fermented on purpose or as a result of hit-or-miss storage is something people can argue about forever. What's hard to dispute is that the taste of fermented meats is well loved around the world.

To ferment meat is to slow the deterioration of the flesh by making it difficult for dangerous bacteria, yeasts, and molds to grow. Fermenting meat is a great way to take an ordinary piece of meat, say beef for example, and turn it into an entirely different flavor, with little more than some spices and time. It's a tried-and-true method of expanding the palate, while using the foods you have available.

Fermenting meat makes it tender and easier to digest. Because many fermented meats start with less than choice cuts, the mere act of fermentation transforms a tough, chewy bite into a luxurious, tender one, with a memorable flavor. Think of fermented meats as the much-loved answer to frugal cooking. You can turn the less expensive cuts of meat into the main dish at any meal with a few simple ingredients.

In this chapter, we show you types of fermentation that involve meat, fish, and eggs.

Choosing Meat and Ingredients for Fermentation

Choosing meats, fish, and eggs for fermenting means recognizing and choosing only the freshest cuts of meat, the freshest fish, and eggs that have been purchased soon after having been laid.

This goes back to sourcing your food, and getting to know the farmer who raised or sold you the product.

When buying meat from a butcher, let him or her know that you will be fermenting a specific cut of meat. He will keep the piece a manageable size and not potentially grind it into burger as he would normally.

Choosing your meats is also common sense. If you can't stand the taste of spicy sausage, you probably don't want to make 20 pounds of homemade spicy sausage. Choose foods that you like to eat from the store, and try to find recipes for making them at home.

Selecting spices, herbs, and flavorings

After you know what types of fermenting you will be doing, it is time to get creative and choose your add-ins. For me, this means going easy on the red pepper and increasing some of the herbal parts. I can't handle very hot foods, so fermenting my own and cutting down on the heat allows me and my family to enjoy a wider variety of previously off-limits foods.

Like any good recipe, your end results will only be as good as the ingredients you put in. Choose fresh herbs and spices. Use flavoring suggested in a recipe to begin with, changing it up once you understand how the fermenting process works and how the recipe is supposed to taste.

A fermented recipe is no place to use up dried out, flavorless herbs and spices. If you can't easily smell your add-ins when crushed, then they are too old. Find something fresher or leave them out altogether.

We all have different palates, but there are some basic flavors that go into most fermented meat recipes:

- ✔ **Kefir:** A fermented product, kefir is created with bacterial grains that resemble cauliflower. These gelatinous grains are removed and reused over again. They eventually grow in size and can be divided to share.

✔ **Nitrates or nitrites:** This ingredient is used for flavor enhancement, provides the appealing pink-cured color, and helps prevent rancidity. We talk more about choosing to add nitrates or nitrites later in this chapter.

✔ **Salt:** Flavors the meat, but also draws out moisture from within the meat, making it less inhabitable for bacteria that can cause illness. Salt also helps the cure to penetrate the meat, bringing the flavor throughout the entire thickness of the piece. Although you can use salt on its own, it would be unpalatable to eat, and the meat would take on an unsavory color.

✔ **Spices:** Herbs and spices are added to the brine to create a unique recipe. Spices are the single thing that can change a recipe from boring to taking center stage. Add these flavors to make a less than palatable cut of meat taste appealing. This is the perfect place to start playing with heat, too. The combination of salty and hot is a surefire hit for most people. Look for powdered hot peppers and make your meat as spicy as you dare.

✔ **Sugar:** To counteract the salty taste and add a new facet of flavor, sweetener is sometimes added to the curing process. Most commonly used in wet brine (which is fermenting), sugars can include cane sugar, honey, or maple syrup. Each brings its own flavor to the recipe.

It may come as a surprise to you that an essential ingredient in most brines is the sugar. The sugar can be white, brown, maple syrup, honey, molasses, or any combination of sweeteners. These unexpected flavors add an additional flavor component, smoothing out the saltiness and helping to take the sting out of the spiciness factor. You can't ruin the effectiveness of a recipe with the sugar ingredient, so feel free to play around with this ingredient to your heart's content.

Selecting starters

When fermenting meats, fish, and eggs, it is acceptable to start from scratch, and your recipe will turn out very well. Fermenting hobbyists will often add a bit of old fermented liquid to the new recipe, to give the bacteria a good boost right from the beginning. This is as simple as it sounds: Add a little bit of starter from a successful batch of fermented food to the new jar or container of the same food. It's that easy!

Although not necessary for all ferments, I (Amy) highly recommend using starter from a previous batch of the same recipe, whenever possible. Think of it as insurance for a healthy, new ferment.

Nitrates or not?

As food preservation developed over time, people realized that meats were more flavorful and more visually appealing with the addition of potassium nitrate. This chemical, also known as saltpeter, kept the meat an attractive pink color and helped to inhibit the growth of bacteria. It gave preserved meats the tanginess you associate with these foods. Although people no longer use saltpeter, sodium nitrite, the modern version, is still used.

There has been much controversy over the use of nitrates and food safety. The U.S. Department of Agriculture (USDA) had come to the conclusion that the amounts of nitrates in meat should be regulated, but under its guidelines, nitrates are considered safe.

Nitrates are used in such small amounts that they are mixed with salt in order to be accurately measured. For this reason, preserving mixes are often premade and sold with the nitrates included. Using a mix ensures that you don't use too much. This ingredient is sold in premixed form under names like Morton's Sugar Cure, Tender Quick, or Bradley's Sugar Cure. You add these mixtures to meat based on weight. What could be easier?

When I (Amy) am making fermented meat for my family to eat, I often leave out the pink salt. I do this for a few reasons:

✔ We are going to be eating the small batch of meat (less than 5 pounds) in one or two meals.

✔ We don't need the pinkness in order to enjoy the taste.

✔ We eat our fermented meat more often than the USDA-recommended 1 time a week, and therefore, our consumption of nitrites would be higher than the recommended dose over time.

Choosing Casings

One popular form of fermented meat is sausage. When many of us hear the word sausage, the types of meats that come to mind often have been stuffed into a casing. Casings are the wrapping that holds the sausage shape. Casings can be made from many things, from removable tin foil to animal intestines. This casing holds the meat and other ingredients together and is for visual appeal in many recipes, but it can also be an essential element in some types of sausage recipes. Remember to check your recipes to see if you are going to need casings before you start.

Natural casings

Casings and wrappings were once only one thing; they came from the cleaned and sanitized intestine from a sheep or pig. For thousands of years, this was the only type of casing available, and they are still used today. These natural casings are what provide the bite, or snap, that some sausages have. Natural casings are completely edible, although they may be difficult and expensive to buy for the home sausage maker.

Artificial casings

Casings can also be made from other substances. Casings made from collagen are more readily available and less expensive. They come in both edible and inedible varieties, so note the difference on the packaging. They've been improved in recent years, providing a more flavorful bite than they were once known for. Collagen casings are easier for the home sausage maker to manipulate, making them a good choice. Made from the natural collagen found in beef or pork hides, tendon, and bones, they are completely digestible when noted as such on the packaging.

Casings can also be made from completely inedible ingredients, such as plastic or cellulose. Cellulose casings are designed to hold the sausage shape through the cooking and/or curing process and then are made to be cut or peeled away. Modern mass-produced hot dogs or skinless franks are made by using cellulose casings.

You can also use oven wrap-style plastic, parchment paper, and even foil in sausage making. For the home sausage maker, these materials may be the most useful way to experiment with sausage links. These DIY casings are perfect for some of the fancier sausages that you'll be making tiny batches of, when pulling out the stuffer is more work than it's worth. Making your own wrapping is also a great way to jump in and get started without investing much time and money.

Look for the recommended width of the casings that a recipe calls for. They come in various sizes, from ¾ inch to 20 inches in diameter!

Meat Fermentation Techniques

Making fermented meats is much the same as fermenting any other foods; you must adhere to strict sanitary guidelines and watch for many of the same issues that you do with other preserved foods. To muddy the waters a bit, there are different terms used for the same techniques, depending on where you are. For instance, you may call making corned beef curing or corning. It is also known as fermenting, since the meat sits in wet brine (anaerobic environment) for days, while the bacteria work to break down the proteins in the meat, making it tender and delicious. It is always a good idea to discern the actual method of meat preservation, rather than just going by the term used. This ensures that you are making the process perfectly clear.

Fermenting, also called curing or wet brining, is the method that requires the meat to be submerged under liquid, creating an anaerobic environment, where the lacto bacteria can work its magic. Dry rubs applied to meat and left for a time before cooking are also called curing. This is not the same technique.

Pickling is sometimes confused with fermenting. Pickling a meat is adding vinegar to the solution covering the meat, which tenderizes it and changes the flavor — much like fermenting, but pickling is not fermenting, and pickling meats still requires them to be pressure canned to render them safe for storage. There are some quick pickling methods, but these are not meant for storage, and the food must remain refrigerated at all times.

When fermenting fragile foods like meat, fish, and eggs, always use the utmost care with cleanliness, measurements, and modern recipes. It is fine to use an older recipe as long as you can find an updated version with modern safety guidelines.

Grinding and mixing meat

When making sausage, careful attention must be given to the process, since the very nature of sausage making requires many steps that could introduce unwanted bacteria to the mix.

Grinding and mixing meats with other ingredients is what makes sausage making such an exciting hobby. The varieties are endless, once the basic recipe is completed. Find out how to choose the best quality ingredients and the different ways to prepare your meat.

Getting your grind on

Meat must be ground in order to make sausage. When grinding meat, the more spiced the sausage, the coarser the grind you need. Most grinders come with blades (plates) that chop the meat into the recommended grind. Follow the recommendations in the recipe.

To get the right texture, you may need to grind the meat before changing the plate and regrinding. Water and/or ice is often added to the meat because grinding causes heat, which melts the fat. Keeping the meat and fat cool as you grind is necessary for a desirable texture.

Mixing your sausage meat is the next step to making sausage. Your ingredients should remain cold and be fresh. Have a plan for processing before you start, your prep should be finished and you have an idea of how to proceed, in order for the ingredients to remain the proper temperature.

Here are some tips for mixing meats safely:

✔ Reduce the spread of bacteria by washing your hands before and during the sausage-making process.

✔ Start with a clean work area and clean meat. This means fresh meat, kept at the proper temperature.

✔ Marinate raw meat in the refrigerator only. Do not allow it to come to room temperature.

✔ Use a food thermometer as you work to be sure of the temperature of the ingredients. Do not rely on guessing or simply feeling the raw meat with your hand.

After your meat is combined with additional ingredients, it is placed in a refrigerator to give the flavors time to develop. This is part of the fermentation process and is essential for specific types of sausages. Another fermentation step is in the hanging of certain types of sausage. In fact, some sausages have visible mold.

Choosing meats

Making sausage starts with the meat itself. Choose meat with no skin, gristle, bone, or blood clots. It goes without saying that fresh meat makes the best fermented meat and will result in a better taste.

No one meat makes the best sausage; each variety has its fans. Meat should be high quality and have the right lean meat to fat ratio. It should also be clean. This means from a healthy animal, packaged in a sanitary environment, ground in a clean machine, and handled every step of the way, using proper sanitary guidelines.

Mixing in fat and other ingredients

Mixing meat together with fat is essential for a good-tasting sausage. Like the meat, a good-tasting fat is important to your recipe and can ruin or improve the overall taste of the final product. Flavorings like salts, herbs, and fillers all add to the dimension of taste that you will find in the world of sausages. You'll find a wide variety of old-school ingredients to try, with plenty of new ideas for flavorings. Sausage making is a never-ending hobby, with endless variations.

Fillers like bread, dry milk, flour, egg, bulgur wheat, onions, and garlic do more than extend the recipe size; they are also important for flavor and texture. When fermenting sausage, care must be given because of the additional opportunity for contamination. The additional fillers also shorten the shelf life of sausage and should be considered for sausages that you plan on using right away.

Only add fillers if a recipe gives instructions to do so. Adding fillers to a fermented recipe without guidance can tip the scales in favor of dangerous bacteria and yeasts, making it easier for them to grow.

Stuffing sausages

Once your meat is ground and ingredients mixed, it is time to stuff your sausage casings. Although not always essential, stuffing sausage mixture into a casing makes them the familiar shape and texture that we all know. Of all the steps, the actual stuffing is the only one that takes some getting used to. It isn't difficult, but it takes some dexterity and timing to get the links a uniform length.

To stuff sausages, you do not need a stuffing attachment. You can simply make patties or leave the sausage loose, and package by weight. Alternatively, you can create a temporary casing for cooking the sausage, out of inedible material like foil or parchment. This would be removed when the sausage was finished cooking, and works very well for someone new to making homemade sausages, without much investment in supplies.

Brining

Brining is the term used when your meat is submerged into a liquid that usually is seasoned with salt, sugar, and spices. The key to brining is to keep the meat under the liquid, where the good anaerobic bacteria can then work on the meat. This can also be defined as fermentation.

Curing

Curing meat can mean different things. In the case of fermenting, curing is the time you allow the meat mixture to sit in the refrigerator, before eating or freezing. This cure time is also the fermenting stage.

Smoking

Smoking meat is a process in which the meat is subjected to either hot or cold smoke, to infuse with the flavor of the smoke. Smoking meat is an additional step in the preservation of meat and can be done to many types of fermented meats as a final finish.

Making Food Safety a Priority

The method for creating delicious and safe fermented meats, fish, and eggs is simple:

- ✔ Keep everything the correct temperature: cold things stay cold and go directly to the stove still cold. Hot foods come up to temperature and are eaten hot.

- ✔ Know your steps and get the product from the recipe to the fridge as efficiently as you can.

When fermenting something as delicate as meat, eggs, and fish, there are clear and not-so-clear signs that something has gone wrong. Here are some of the ones to watch out for:

- ✔ **Bulging casings:** This issue may be caused by air pockets that remain after the initial filling of the casings, so when you do the filling, poke air pockets out as you go and work carefully. If you see any new bulges, however, discard the sausage.

- ✔ **Forgotten meat or meat that's left out too long:** It happens. Everyone forgets a batch, or something gets pushed back to the rear of the fridge. If your sausage has been forgotten and comes to room temperature, or if you discover it in the back of the refrigerator, discard it. Even a tiny amount of bad bacteria can make you seriously ill.

- ✔ **Mold:** Molds are tricky. For the beginning sausage maker, mold should never be present. Some molds help create specific types of sausage (salami for example), but they're fuzzy white molds, and the experienced sausage maker will know to look for it during a specific time frame. At home, any mold is suspicious, especially if it's blue, green, or black in color. Discard any moldy meat.

- ✔ **Off smells:** Distinguishing these odors in cold sausage may be difficult, but do smell the product carefully before eating. Sometimes you miss it in the cold product, but as it begins cooking, the off odor becomes very clear. Discard any sausage that you think even *might* smell bad.

Choosing a Spot to Ferment Meat

Fermenting requires time for the food to actually ferment. This means using a space that can be filled for quite some time, possibly with large bowls and containers. A shelf in your refrigerator will work very well, as long as you can give it up for a week. Many hobbyists purchase a second refrigerator for just this thing. That way the family refrigerator stays organized with everyday foods.

Determining space needs

Like any slow or homemade food, fermenting can take up quite a bit of space. You will be using multiple containers and revolving them constantly as one fermented recipe finishes and the next gets started fermenting. Try to find an out-of-the way storage area that will not interfere with your day-to-day cooking.

When fermenting large cuts of meat, it is acceptable to cut them down to fit into smaller containers. It will not interfere with the fermentation process.

Controlling the environment

Fermenting means knowing all the variables. Your fermenting temperature for delicate foods like meat, fish, and eggs needs to be between 35 and 40 degrees (the normal range of a refrigerator). Keep the container tightly sealed during the curing stage, and the flavors of other strong foods won't interfere with your fermenting recipe.

Finally, keep your ferments in clean, sanitized containers in the refrigerator. Use clean hands to check the recipe, and always have your plan in place before working with the food items. That way, you reduce the time that meat can sit and come to room temperature.

Should you get your own fermenting chamber?

As you become more proficient in home fermented sausage making, it might be time to move on to dry cure fermented sausage recipes. These sausages, like prosciutto and salami, need weeks and months to cure, so having a fermenting chamber is the next logical step.

For the general hobbyist, however, a fermenting chamber is not necessary, and the short time frame that the beginning recipes in this book need can be completed right in your refrigerator.

Storing Fermented Meats

After your meats are fermented and taste just the way you like them, it is time to consider storage. Fermented meat, fish, and eggs must remain under the proper conditions to stay safe. This means eating within a day or two of finishing the recipe, or storing safely.

The safest and most efficient way to store fermented foods is to freeze them. Ideally, use heavy freezer bags to seal the meat completely and remove all the air. This keeps your sausages tasting fresh and frees up space for you to try your next fermentation experiment.

The next best thing is an airtight container with a lid, or wrapping in plastic wrap and then tightly wrapping in freezer paper.

Corned Beef

Prep time: 20 min • **Cook time:** 2 hrs 30 min • **Ferment time:** 5 days • **Yield:** About 5 pounds

Ingredients	*Directions*
1 gallon water	**1** Combine the water, salt, sugar, pink salt, and 2 tablespoons pickling spice in a large pot. Bring to a boil, stirring often, until the salt and sugar are dissolved.
2 cups kosher salt	
½ cup sugar	
4 teaspoons pink salt	**2** Remove from heat and cool to room temperature. Refrigerate until cold.
4 tablespoons pickling spice	
5 pounds brisket	**3** Place the brisket in a large stockpot and cover with the cold brine. Refrigerate for at least 5 days.
2 medium onions	
2 medium carrots	**4** Remove the meat from the brine and rinse. Place the meat in a large pot. Add just enough water to cover the meat and add the 2 remaining tablespoons pickling spice.
2 celery stalks	
6 cloves of garlic	
	5 Chop the onions, carrots, celery, and garlic and add them to the pot. Bring to a boil. Reduce heat and simmer 2½ hours, or until the meat is tender.

Salami

Prep time: 30 min • **Ferment time:** 24 hours • **Cook time:** 1 hour • **Yield:** About 2 pounds

Ingredients	Directions
2 pounds ground beef	**1** In a large bowl, combine all ingredients.
2 tablespoons Morton's Tender Quick Salt	**2** Place into the refrigerator tightly covered, for 24 hours.
1 teaspoon whole peppercorns	
½ teaspoon while mustard seed	**3** Remove from fridge and divide in half. Shape each half into a log and roll in foil. Tightly twist ends.
½ teaspoon red pepper flakes	
½ teaspoon minced garlic	**4** Place the logs in the refrigerator for 1 hour to firm up.
	5 After an hour, place wrapped logs into boiling water and return to a boil. Boil for 1 hour.
	6 Remove from boiling water and unwrap to drain before placing finished salami into the refrigerator. Meat may be refrigerated or frozen.

Venison Sausage

Prep time: 15 min • **Ferment time:** 3 days • **Yield:** 10 pounds

Ingredients	*Directions*
6 pounds ground venison	**1** Combine the ground meats and mix well. Combine the red pepper, salt, sage, pepper, and garlic. Stir the spice mixture into the water.
4 pounds ground pork shoulder	
4 tablespoons crushed red pepper flakes	**2** Pour the spice and water mixture over the ground meat and mix well. Make patties and refrigerate for up to 3 days, or freeze for up to 3 months.
6 tablespoons kosher salt	
1½ tablespoons ground sage	
5 tablespoons black pepper	
3 tablespoons garlic powder	
1½ cups ice water	

Summer Sausage

Prep time: 30 min • **Cook time:** 1½ hr • **Ferment time:** 4 days • **Yield:** Two 1-pound rolls

Ingredients	Directions
2 pounds ground beef	**1** Mix all the ingredients well and place the mixture in the refrigerator for 12 hours. Remove from the refrigerator and remix. Divide the cold mixture in half and shape into rolls.
2 tablespoons Morton's Tender Quick Salt	
1 teaspoon liquid smoke	**2** Wrap each roll tightly with foil. Place the rolls in the refrigerator to cure for up to 4 days. Poke several holes in the foil to allow excess fat to drain away.
½ teaspoon garlic powder	
1 teaspoon black pepper	**3** Place the sausage in a broiler pan and bake at 325 degrees for 1½ hours. You can eat this sausage cold and freeze it for up to 3 months.
½ cup cold water	

Corned Venison

Prep time: 30 min • **Cook time:** 4 hr • **Ferment time:** 3 weeks • **Yield:** 10 pounds

Ingredients	Directions
5½ cups salt 3 gallons water 4 cloves garlic 1 cup brown sugar 3 tablespoons pickling spice 10 pounds venison roast	**1** In an 8-quart saucepan, dissolve the salt into the water. Add the garlic, sugar, and pickling spice to the salted water. Bring the pickling solution to a boil over medium-high heat. Reduce the heat and simmer for 5 minutes. Allow the brine to cool to room temperature.
	2 Place the venison in a container large enough to hold both the meat and the brine. Cover tightly and refrigerate immediately. Allow the venison to pickle for 3 weeks, skimming any foam that develops.
	3 After 3 weeks, remove the meat and place it in a clean, 8-quart saucepan. Add enough fresh water to cover the meat. Bring the water to a boil over high heat. Lower the heat to medium and simmer the meat for 4 hours.

Fermented Salmon

Prep time: 15 min • **Ferment time:** 24 hours • **Yield:** 1 pound

Ingredients	Directions
1 pound salmon fillet, cut into bite-sized pieces	*1* Pack the salmon into a quart jar.
1 cup water	
¼ cup whey	*2* Combine the rest of the ingredients in a small bowl.
1 tablespoon sugar	
1 tablespoon sea salt	*3* Pour the liquid over the fish until it's completely submerged. Add more water if needed.
¼ organic lemon	
¼ cup fresh dill leaves	*4* Cover and leave at room temperature for 24 hours. Place in the refrigerator for up to 2 weeks.
8 whole peppercorns	

Fish Sauce

Prep time: 15 min • **Ferment time:** 3 days plus 3 weeks • **Yield:** 2 cups

Ingredients	*Directions*
1½ pounds fish, heads included	**1** Cut the fish and lemon rind into small pieces and place them into a quart jar. Sprinkle sea salt over all and press down with a wooden utensil.
½ organic lemon rind	
3 tablespoons sea salt	**2** Add the rest of the ingredients to the jar and stir. Add additional water so all ingredients are covered with liquid.
2 cups water	
2 cloves garlic	
2 bay leaves	**3** Cover and leave at room temperature for 3 days.
1 teaspoon whole peppercorn	**4** Place the fermented fish sauce in the refrigerator for 3 weeks.
3 tablespoons organic whey (use the clear liquid that rises to the top of plain, organic yogurt)	**5** Strain the liquid and keep the sauce in a glass container to use. Store it in the refrigerator for up to 3 months.

Lacto-Fermented Eggs

Prep time: 15 min • **Ferment time:** 4 days • **Yield:** 2 cups

Ingredients	*Directions*
12 hard-boiled eggs, peeled **2 tablespoons sea salt** **6 tablespoons organic fresh whey, from another batch of fermented food or plain yogurt** **Water to cover eggs**	*1* Place as many eggs as you can easily fit into a quart jar. You may need two jars for this.
	2 Combine the sea salt, whey, and water. Pour this mix over the eggs until they're fully submerged.
	3 Cover and leave at room temperature 24 hours.
	4 Place in the refrigerator for 3 more days.

Part V
Beer, Wine, and Other Beverages

Visit www.dummies.com/extras/fermenting for more information on kombucha, a wonderful, healthy fermented beverage.

In this part . . .

- ✔ Find out about the health benefits of fermented drinks.
- ✔ Gain an understanding of the fermentation process at every stage.
- ✔ Discover how to choose starters that best work with your tastes.
- ✔ Explore a variety of beer and wine recipes.

Chapter 14

Healing Beverages

In This Chapter

▶ Learning about healthy and healing drinks
▶ Making natural carbonated drinks

*Y*ou can gain plenty of health benefits from your daily drinking! In this chapter you learn all about the wonders of healthy fermented drinks and how to make them easily from the comforts of your own home. From ancient fermented Asian teas like Kombucha to Russian drinks like Kvass, you learn the art of making unique healthy drinks from around the globe. These fizzy-pop-like creations will inspire you to learn new things and discover nature's magic and fill you with healthy fermented flora!

Here are some things each of your fermented drink recipes all have in common:

✔ Fermented drinks naturally become carbonated and fizzy.

✔ Fermented drinks are naturally full of living probiotics.

✔ Fermented drinks use starters like yeast or kefir grains.

✔ Fermented drinks produce trace amounts of alcohol.

✔ The longer your drink ferments and sits out, the more sour, sweet, or vinegary its taste will become.

✔ Sweeten all your fermented drinks as you desire using honey or an alternative sugar.

✔ Be sure when bottling to leave some head space in the top or it could explode from fizzy-pressure.

✔ You can vary your recipes using different fruits or herbs.

Choosing Starters

Starter cultures can be wet or dry, easy to make or hard to find. They are commonly used for dairy, meats, wine, beer, and bread. So what exactly is a starter culture and why do you need it? Take a look at these frequently asked questions around fermented foods and the strange world of cultures. It can be intimidating at first, but once you get the hang of how starter cultures work, you'll be an expert in no time.

The greatest role of the starter culture is its ability to speed up the fermentation process and get guaranteed results. Starter cultures are the microorganisms that inoculate your fermented foods to help get the desired changes we want to see. Any culture starter will give your recipe added enzymes and flavors, change its texture, and add health benefits.

Do all recipes require a starter culture? No. It is true that not all fermented foods require a starter culture. Be sure to check your recipe and see which ones do and do not. Some bread, for example, might use yeast or culture, like for making sourdough bread, while for making sauerkraut, a fermented cabbage recipe has naturally occurring microorganisms that do not require this at all.

Finding starter cultures

Starter cultures can be purchased commercially or made at home. All starters are made up of naturally occurring microorganisms, most notably the beloved *Lactobacilli,* and a combination of other food products such as water and flour or dairy products like milk or yogurt. Depending on your recipe, the starter will be made from various organic items.

In commercial yeasts, such as brewers yeast or active dry yeast, the yeast is carefully selected but can stale easily and needs to be replenished by continuous market purchase. Although it does not always need oxygen it does not thrive in acidic conditions, which are ideal for fermented goods. On the contrary, wild yeasts or homemade starters require oxygen, don't stale, as easily, and are consistently living for as long as we feed them. They are kept alive by feeding them sugars and starches (such as from flour). These wild yeasts give our sourdough a delicious fresh smell and can be tasty and feel quite different from the commercial products offered. Wild yeasts help you save on costs and have the freshness and full nutrients that you desire.

What the heck is backslopping?

Backslopping is a term that refers to the use of some leftovers from one ferment to another. It is the leftover product that is used to kick-start your next batch of fermented goods.

This backslop, or starter culture, helps pump your new recipe with guaranteed results — you don't want to use just any backslop, but use your best backslop!

Using starter cultures

Depending on your recipe, the starter culture will be different. For example, with sourdough bread (see Chapter 8 for a great recipe), your very own homemade starter begins with simply combining water and wheat flour. For others you might use fermented grains, a yeast, or bacteria. No matter what starter culture you decide to use, starter cultures are easy to make. Starter cultures take time to grow, but you can save some time by purchasing starters pre-made from friends, at health food shops, or online. The beauty is that once you master making them, you can keep them for a long time and continue to use them to make more of your favorite fermented cultured foods.

Understanding the differences between alcoholic and nonalcoholic brews

Though it may seem intuitive, there is actually quite a spectrum and scale from nonalcoholic to alcoholic drinks. There are drinks made with no alcohol, low alcohol, mid-range, and strong amounts. Some alcohols can be carbonated, while others are distilled. Some alcohols are made from fruit juice and turn into vinegars, while others are made into grains and turn into beverages.

The most common fermented drinks that contain a high percentage of alcohol are wine, beer, and spirits. These drinks are created with the magic combination of water, yeasts, and sugars to get the desired product. For more great recipes on these mood-altering drinks, see Chapters 15, 16, and 17 and learn how to make your own alcoholic homebrews.

The fermented drinks you learn to make in this chapter contain trace amounts of alcohol. The amount of alcohol content is extremely small and will not act on your body the way an alcoholic beverage will. In fact, you may instead experience consequential health benefits, and these drinks typically only contain 0 to 1 percent of alcohol content.

It is easy to make nonalcoholic beverages from the comfort of your home by avoiding the fermentation process and stopping it in the early stage before alcohol develops. Homemade nonalcoholic beverages can have similar tastes and flavors by using fresh ingredients that mock the desired tastes. Nonalcoholic beverages produced in commercial environments can also use this method and avoid the fermentation process or they can also remove the alcohol after the fermentation stage. Just like any food or drink, the amount of alcohol you decide to consume will depend on how long it has fermented for and how much of it you are consuming!

Making Natural Carbonated Drinks

It's hard to beat the satisfaction of making homemade carbonated drinks! You'll learn to love the fresh flavors and soda-like qualities. Making natural carbonated drinks at home is simple and easy to do, with the bonus of being healthy. Take a look at some of the benefits:

- ✔ Homemade carbonated drinks are easy to make.
- ✔ They are much more inexpensive, so you'll save on costs.
- ✔ They're a great replacement for pop or conventional soda.
- ✔ Fresh ingredients make for fresh flavors.
- ✔ You can control the amount of sugar you put in.
- ✔ You can have fun watching your drink bubble overtime.

Ginger soda

There is nothing like a fresh bubbly ginger brew to wake up your senses! Ginger soda has several different steps but it's definitely worth the wait. There are three different stages that the soda goes through. First you will learn how to make the starter, then we will teach you how to make the *wort*, and lastly you will bottle your soda to store in your fridge!

"What is wort?" you might be asking. A wort in homebrewing is the name for the beverage, or soda mix, before you have placed your starter into it and fermentation takes place. It's a term that homebrewers use and is a basic ingredient in homebrewing.

Lacto-lemonade

The joys of homemade lemonade have been tasted and enjoyed around the world. Lacto-fermented lemonade is like regular lemonade, but with the added health benefits of living microbiology that helps us to stay healthy. For all you lemon-lovers out there, this is an easy-to-make drink that's tangy, and nice and refreshing on a hot summer day or a morning pick me up.

Beet, apple, and ginger kvass

This fermented beverage has its roots in Russian pride. It began as brewed drink made from rye bread or beets and has similarities in flavor to root beer or cola. Homemade Kvass is hands down much healthier for the teeth and body since it has no additives, much less sugar, and is full of fermented benefits. Although this particular recipe does not require a starter culture, traditionally yeast was used.

Root beer

Did you know that traditionally root beer was made using a tree called Sassafras? Today, many people use wintergreen and licorice root. Homemade root beer will give you a taste for the good ol' days and is full of herbal infusions.

Kefir

We've said it before and we will say it again, fermented drinks are great for your health and full of probiotics. Kefir is said to originate in Eastern Europe and was a drink that originally began as a way to preserve milk. Kefir comes in all sorts, both dairy and non-dairy kefir alternatives. We will focus on two great non-dairy kefir recipes with a base of coconut and water. The sugar and bacteria in these liquids will create a bubbly and refreshing drink that is a bit sour, but great for the gut. Kefir is an inexpensive drink that can easily be made from the comforts of your own home.

Once you make any starter, like kefir grains from your completed recipes, you can use them again. You can experiment with new flavors and larger batches — this is the beauty of starter cultures and homemade small-batch fermentations!

Kefir grains love and thrive on lots of minerals. If your grains are not fermenting, meaning you aren't getting the fizz or carbonation that you desire, add in 1 teaspoon of blackstrap molasses. This will help your little grains to soak up the sugars and continue to grow.

Amasake

Are you ready to try something new and sweet in your life?

Amasake is an ancient fermented rice drink that is sweet and fragrant. This drink has its traditional roots in Japan where production of rice is abundant. Koji, or fermented soybeans, is essential to this drink, and you can learn to make it yourself or purchase it from your local Japanese food shop.

Koji is a culture that is prevalent in Japan and Japanese foods. Some of the most popular foods that contain Koji are soy sauces, miso, and sake. The koji is responsible for breaking down the carbohydrates and sugars in these fermented food products and is the result of fermented rice or soybeans. It is great in recipes such as amasake! In essence, koji is cultured soybeans or rice that has already been fermented. It can be purchased from Japanese health shops or homemade.

Kombucha

Kombucha has its roots in ancient Asian tradition but has become highly popular. Today you can find kombucha in every health food shop, at yoga studios, amongst new age crowds, and brewing in the crafty corners of exploratory breweries, but more and more in the mainstream. Why the sudden popularity? Likely, it is due to the health properties that captured the hearts and health of many.

Kombucha has a tangy, slight vinegary taste, but its base relies on unrefined sugar, water, and tea bags. It is said to have many detoxifying qualities and to give vitality; sometimes it is known as the *immortality elixir*. Though it is important to note that the base of this drink is unrefined sugar, which regardless of the other health benefits can still be bad in large quantities, in small doses this elixir is full of gut-healing benefits. It is said to increase digestibility, energy, and gut flora.

In order to make kombucha, you need a SCOBY, which stands for the Symbiotic Colony of Bacteria and Yeast. It is essential for kombucha making! Sometimes you will hear people use funny names for it like "the mother", because it is the source of life for kombucha, or a "mushroom" because it turns into a big, gelatinous, and bacterial component that can resemble a fungus. It is normal to see a wide range of SCOBY variations; from bumps, holes, or small stringy remains from the SCOBY culture, it might even float sideways from time to time. The main thing to look out for is any mold. Mold is rare, but dangerous. If you see any signs of blue or black mold on your SCOBY, throw away the SCOBY and the kombucha. You can use the same container again, but just ensure you clean it out well!

Here are some tips to keep in mind when making kombucha:

- ✔ Plain white, black, or green tea work for kombucha.

- ✔ Find a SCOBY that is fresh to get your best results.

- ✔ Choose a wide-mouth jar. Your SCOBY will grow big and form to your jar! With every new batch you make, you will see your SCOBY reproduce and get thicker. It will make SCOBY babies.

- ✔ Avoid brewing your kombucha for too long, but leave it long enough so it turns from too sweet to just sour enough.

- ✔ Although refined white sugar is the best for kombucha to digest, try to use cane sugar or unrefined for better health.

- ✔ Avoid plastic or metal objects to make it! Kombucha needs glass and wood as it becomes highly acidic.

- ✔ If you see any mold on your SCOBY, or if it has turned black, throw the entire thing out!

Ginger Soda

Prep time: 5 min • **Ferment time:** 48–72 hr • **Yield:** 1 gallon

Ingredients	*Directions*
Filtered water	Make the starter for your soda:
Ginger root	*1* Fill a 4-cup jar with water, about two-thirds of the way full.
Unrefined sugar like cane or Sucanat	*2* Grate the ginger, enough for one heaping tablespoon. Do not remove the skin. You will also need ⅓ cup of ginger later for the wort; feel free to do that now.
⅓ cup lemon juice	*3* Place the tablespoon of the grated ginger into the jar of water.
	4 Stir in one large tablespoon of sugar and lemon juice until it is dissolved.
	5 Place cheesecloth and a rubber band on your jar — you want to allow some breathing room out but prevent bacteria or dust from coming in.
	6 Place the jar in a warm place. The ideal temperature is 75 to 85 degrees for the jar to ferment.
	7 Insulate and cover the jar. (You can use cloth towels to surround and cover it.)
	8 Allow the jar to rest and work its fermentation magic!
	9 Stir your starter once a day for 3 to 7 days. Taste and add ginger and sugar as desired to taste! Once you begin to see bubbling, the starter is nearing its finish.
	Make the wort or pre-fermented liquid. This is the part of the recipe that sweetens up your soda!

1   Fill a gallon jar with approximately 10 cups of water, or two thirds of the way full.

2   Add ⅓ cup of grated ginger to the jar of water.

3   Add in 1½ cups sugar, stir well!

4   Take 1 cup from your starter recipe and stir it into the jar.

5   Stir in the remaining ⅓ cup of lemon juice.

6   Stir again until all the sugar is dissolved.

7   Cover your jar with cheesecloth and a rubber band.

8   Let your jar sit in a warm location, or cover with towels in room temperature. Let your wort rest.

9   Stir once or twice a day for 3 to 7 days. Once you achieve the soda taste you desire, bottle or store your homemade soda in the fridge!

Tip: Taste your wort once in a while. Is it sweet and gingery as you desire? Add water, sugar, or ginger as desired. Is your soda fizzing? Add more starter if you need to!

Tip: Always buy fresh ingredients where you can. Ginger root fresh from the market will give you the best flavors for this soda recipe.

Lacto-Lemonade

Prep time: 10 min • **Ferment time:** 48–72 hr • **Yield:** 3 liters

Ingredients	Directions
2 cups lemon juice	*1* Pour your lemon juice, sugar, and kefir grains into a 3-liter jar.
1 cup alternative and unrefined cane sugar like Sucanat or coconut sugar	*2* Stir all your ingredients together until the majority of your sugar is dissolved.
¼ cup kefir grains	*3* Place the lid on the jar and set in room temperature.
2.5 liters filtered water	*4* Watch your drink become carbonated in just a few days!
	5 Taste, refrigerate, and sweeten as desired!

Beet, Apple, and Ginger Kvass

Prep time: 10 min • **Ferment time:** 48 hr • **Yield:** 1 liter

Ingredients	*Directions*
¼ **cup mixed chopped beets and apples**	*1* Place your mixed chopped (not grated) beets, apples, ginger and honey into the jar.
Fresh ginger, sliced	
1 tablespoon unpasteurized or raw honey	*2* Fill your jar with water, leaving ½ to 1 inch of space for air space at the top. The kvass will naturally carbonate and place pressure on the jar lid.
Filtered water	
	3 Seal your jar lid tightly and store your kvass at room temperature.
	4 Leave your kvass to carbonate for 2 to 3 days. Shake and stir daily. You should be able to see the fizzing and bubbling after just 24 hours!

Tip: Taste your kvass and see if it needs more sweetness! Add sugar accordingly. Strain, bottle, and refrigerate for up to one week.

Root Beer

Prep time: 25 min • **Ferment time:** 48–72 hr • **Cook time:** 30–40 min • **Yield:** approx. 18 cups

Ingredients	*Directions*
8 cups water	*1* Bring 8 cups of water to a boil and stir in the sassafras and vanilla bean. Let the pot stay uncovered.
20 inches sassafras root	
1 piece vanilla bean, 3 inches long	*2* Reduce the heat to simmer the roots for 15 to 20 minutes. You will notice the water begin to turn red in color. It should also smell delicious! Smell and color not strong enough? Add more roots!
1¾ cup brown sugar or Sucanat	
⅛ teaspoon ale yeast	*3* Stir in the unrefined sugar until it dissolves.
¼ cup lukewarm water	*4* Let the liquid cool to room temperature.
	5 Drain the roots and vanilla bean through a cheesecloth or thin-meshed strainer, placing the infused liquid into a large jar or pitcher.
	6 In a separate cup combine the yeast and ¼ cup of lukewarm water. Let it sit about 5 minutes.
	7 Add the yeast-liquid to the jar. Close the jar and shake it around to get the yeast going throughout the liquid.
	8 Bottle your root beer! (Best to use flip-top bottles.)
	9 Store at room temperature, or in a cool dark room.
	10 Watch your homemade root beer bubble and fizz over 2 to 3 days.
	11 Once you have enough carbonation in your drink and are satisfied, refrigerate!

Vary It! Try adding other interesting roots and herbs in your root beer recipe for added flavors. Some people like to try licorice root, ginger root, and juniper berry, some wintergreen for some minty freshness, or even wild cherry!

Water Kefir

Prep time: 5 min • **Ferment time:** 48–72 hr • **Yield:** 1 liter

Ingredients	Directions
1 liter water **¼ cup cane sugar, unrefined or raw sugar** **2–3 teaspoons water kefir grains**	**1** Pour the water into your jar, leaving approx. ½ to 1 inch of space from the roof of your jar to the water. You want to leave some head space as pressure will build from natural carbonation. **2** Add ¼ cup of your unrefined sugar until it is dissolved. The water you use can be warm at first to dissolve your sugar, but it should be cool by the time you place your kefir grains inside. **3** Once your sugar is dissolved and your water is at a cool temperature, go ahead and add 2 teaspoons of water kefir grains. **4** Check that there is approx. 1 inch of air space left in your jar. **5** Place the lid of your jar loosely screwed onto the top, or secure your cheesecloth on with a rubber band. **6** The kefir will naturally carbonate and expand the jar lid. Allow your drink to sit at room temperature for 24 to 48 hours. **7** Check on your carbonated water kefir and taste it! Add sweetness as desired. **8** Store in your fridge after a few days, and be sure to seal the lid more tightly but with enough space for some carbonation.

Tip: Once your kefir is bottled or in its final stages, see if it is fizzy enough for your satisfaction. Not fizzy enough? Just add a teaspoon of sugar or molasses to gain more carbonation!

Vary It! Try adding something new. Some people love to spice kefir with ginger, lemon, or fresh fruit for an added refreshing kick! You can also make it with coconut water! Just replace the water with coconut water for a different flavor and a boost of electrolytes!

Coconut Kefir

Prep time: 5 min • **Ferment time:** 48–72 hr • **Yield:** 1 liter

Ingredients	Directions
2–4 tablespoons or ¼ cup kefir grains	**1** Place your kefir grains in a glass jar and cover it all with the coconut milk.
2 cans unsweetened, undiluted coconut milk	**2** Mix together with a plastic or wooden spoon.
	3 Cover your glass jar with a towel or cloth napkin and let the magic work from 12 to 36 hours.
	4 Be sure to check the coconut kefir every few hours and remember to always remove the kefir grains as soon as it reaches the texture and taste you desire.
	5 Close your glass jar with the lid and place your delicious kefir in the fridge.

Tip: If kefir is left to sit too long it becomes sour and thick! Remember to remove the kefir grains when your drink has reached its desired flavor. You'll notice the kefir thicker in the fridge, but this is a normal process!

Chocolate-Mint Coconut Kefir

Prep time: 5 min • **Ferment time:** 48–72 hr • **Yield:** 1 liter

Ingredients	Directions
2 cups coconut kefir (see preceding recipe)	*1* With your coconut kefir all ready to go, add in cacao, maple syrup, and mint essential oil.
2 tablespoons cacao powder	*2* Make sure all the ingredients are well combined.
1 tablespoon maple syrup	
1 drop mint essential oil	*3* Cover your glass jar with a towel or cloth napkin and let the magic work from 12 to 36 hours.
	4 Be sure to check the coconut kefir every few hours and remember to always remove the kefir grains as soon as it reaches the texture and taste you desire.
	5 Close your glass jar with the lid and place your delicious kefir in the fridge.

Vanilla Cinnamon Coconut Kefir

Prep time: 15 min • **Ferment time:** 48–72 hrs • **Yield:** 2 cups

Ingredients	Directions
2 cups coconut kefir	**1** Mix all of your ingredients together into a jar.
½ teaspoon fresh vanilla bean powder	**2** Cover your glass jar with lid, a towel or cloth napkin and let the magic work from 12 to 36 hours.
1 tablespoon honey or maple syrup	**3** Be sure to check the coconut kefir every few hours and remember to always remove the kefir grains as soon as it reaches the texture and taste you desire.
¼ teaspoon cinnamon	**4** Close your glass jar with the lid and place your delicious kefir in the fridge.

Amasake

Prep time: 10 min • **Cook time:** 20 min • **Ferment time:** 16 hr

Ingredients	*Directions*
White or brown rice (sweet brown rice preferred) ***Koji (see recipe on next page)** **2 cups water**	**1** Cook your rice in a regular fashion and according to the package.
	2 Remove your rice, and let it cool down until it is warm or at room temperature.
	3 Add and stir in 2 cups of koji.
	4 Cover the rice well so it does not dry out.
	5 Leave it in a warm incubator space for up to 16 hours. Keep the mixture at 130 degrees inside of your incubator.
	6 Place 1.5 cups of the sweet rice and 1 cup of boiling water in a blender for a nice, sweet warm beverage!

Koji

Prep time: 10 min • **Cook time:** 20 min • **Ferment time:** 48–60 hr

Ingredients	*Directions*
Sweet brown rice (brown over white rice is a healthy option)	*1* You always want to rinse your rice with cold running water until it no longer runs a strong milky color. This removes all the starch!
Koji culture (known as koji-kin, you can find this in Japanese stores or order it online)	*2* Leave the rinsed and drained rice to sit for up to 12 hours or over night.
	3 Drain any excess water from the rice using a strainer and leave for 2 to 4 more hours.
	4 Slowly steam the rice until it has softened. You want your rice to be slightly sticky, and be sure not to boil it. Keep your lid on the steamer to get great results.
	5 When the rice has cooled down to room temperature, mix in about ½ teaspoon of your koji culture. (Remember to use a wooden or plastic spoon!)
	6 Place the koji-inoculated rice in a shallow tray.
	7 Cover with a damp cloth to prevent it from getting too dry.
	8 Leave your rice in the chosen incubator for up to 40 hours. Keep the temperature of your koji-inoculated rice at around 100 degrees and high humidity.

9 Check on your rice culture from time to time. You'll slowly notice small white fibers develop on your rice — this type of mold is a normal part of the fermentation process! You may even notice a sweet or cheese-like smell. If you recognize bad molds or smells, do not continue, but try this recipe again!

10 Mix the rice 2 to 3 times during the incubation period. After 7 to 8 hours the temperature of the rice will naturally raise itself!

11 Avoid the temperature rising beyond 111 degrees. Take the temperature from the middle of the rice!

12 Remove the rice from the incubator. You can keep koji in a fridge for 1 to 2 weeks or freeze it!

Tip: Do not use boiled rice for koji! Boiling or simmering rice ends up making very slimy and gooey rice instead.

Green Tea Kombucha

Prep time: 20 min • **Ferment time:** 5–10 days • **Yield:** 12 cups

Ingredients	Directions
6 green tea bags	**1** In a large pot, combine the tea bags, sugar, and 2 to 3 cups of water.
1 cup sweetener of choice, preferably unrefined sugar	**2** Bring to a simmer and remove from heat.
3 or 4 liters water	**3** Cover and let the green tea bags steep for 10 to 15 minutes.
A kombucha SCOBY	**4** Add 2 cups of cool water to bring the tea temperature down.
1–2 cups of raw apple cider vinegar or leftover kombucha	**5** Place the SCOBY and finished kombucha or raw apple cider vinegar in your large gallon jar.
	6 Add cooled-down tea to the jar.
	7 Use a wooden spoon (metal is not good) to help the SCOBY float at the top. It's okay if it is sinking, but it should rise during fermentation.
	8 Cover the jar with a paper towel or cheesecloth, and secure with a rubber band.
	9 Let the liquid ferment at room temperature for 5 to 10 days.

10 Taste to see if it is not too sweet and if it is fizzy enough. Sweeten as desired to your taste!

11 When the kombucha is well carbonated and has a slight vinegary smell, pour the kombucha through a strainer into a clean jar or jug.

12 Be sure to leave the SCOBY and approx. 1 or 2 cups of finished kombucha behind. This is your starter for a second batch of tea!

13 Refrigerate it to prevent it from souring any more or if needed, keep it at room temperature for a few days to build up more carbonation.

Tip: Don't let your kombucha build up carbonation too long or it could explode when bottled. Also, the longer kombucha sits out, the more sour and vinegary its taste will become.

Honey Kombucha

Prep time: 20 min • **Ferment time:** 5–10 days • **Yield:** 12 servings

Ingredients	Directions
6 green or black tea bags 12 cups water	*1* Combine the tea bags and a few cups of water in a large pot.
1 cup raw honey or unrefined sugar like Sucanat	*2* Bring the water to a simmer, and remove from heat.
Kombucha SCOBY	*3* Stir in the honey until it is dissolved.
1–2 cups raw apple cider vinegar or leftover kombucha	*4* Cover and let the green tea bags steep for 10 to 15 minutes.
	5 Add 8 cups of cool water to the pot to cool down the tea.
	6 Place the SCOBY and raw apple cider vinegar in your large gallon jar.
	7 Add your now cooled-down tea to the jar.
	8 Use a wooden spoon (metal will harm your bacteria) to help the SCOBY float at the top. If it does not rise, it should rise during fermentation!
	9 Cover the jar with cheesecloth or a paper towel, and secure with a rubber band. Let ferment at room temperature for 5 to 10 days.
	10 Check on it and taste it! Is it tangy, sweet, and fizzy enough?
	11 Pour the kombucha through a strainer into a clean large jar or several bottles. (In this case, leave the SCOBY and 1 to 2 cups of finished kombucha behind for your next batch!)
	12 When you have the taste you desire, refrigerate!

Vary It! Want some new flavors in your kombucha? Kombucha can be made with a wide range of added flavors! Try adding a slice of ginger, or throw a few juniper berries into your next batch!

Chapter 15

Making Wine from Water and Fruit

These are the greatest days for wine consumers in the 8,000 years of wine drinking. Wine quality around the globe has increased exponentially in the past three decades. Good-to-great wine at affordable prices fills the shelves at supermarkets and wine shops, and if you can't find the wine you want down the street, chances are you can find it on the Internet. Today's wine fans have access to a vast range of vintages, regions, and price points.

Winemaking is fermenting grape juice, so if you're making other fermented foods, then making wine is just another step in the world of home fermentation.

What makes a good wine is highly subjective. It helps if you have a favorite type (or types) of wine and start learning what makes that type different from the next. Is it the grape, the length of ferment, or even the soil that the grape is grown in? Making your own wine allows you to create the perfect blend that suits your palate and satisfies that need to create something from start to finish that tastes great!

Did we mention that making your own wine is a ton of fun?

Getting Yourself Ready to Make Wine

When you decide to take the plunge, a defining question quickly arises: What kind of wine are you going to make? We don't mean deciding from among grape varieties; we mean, are you doing this to save money on wine or are you shooting for home wine good enough to compete with the pros?

Health benefits of wine

To be clear, when referring to the health benefits of drinking wine, moderate consumption is one to two four-ounce glasses a day. So we certainly don't recommend more than that, but the benefits can't be denied. Some benefits include:

✔ Lowering the risk of heart disease

✔ Reducing the risk of type 2 diabetes

✔ Lowering mortality rate

✔ Reducing the risk of stroke

✔ Slowing the decline of brain function

✔ Reducing the risk of colon cancer

✔ Lowering the risk of cataracts

Saving money on your adult beverage expenditures and making premium quality wine aren't mutually exclusive; many home winemakers do both. But these two divergent paths present themselves again and again, in the many small decisions you make. Some home winemakers want to economize; some want to prove they can keep up with the big players. In a sense, you'll be training yourself to be a wine connoisseur. Don't worry; this is a highly subjective position. What tastes fabulous to you may not be tasty to the next person.

That's what makes winemaking such a wonderful hobby. You get to know everything about the humble grape and end up with a pretty delicious beverage at the end of the process.

Getting supplies

Before you start working with grapes, you need to think through all the equipment and supplies you'll need and plan on whether to buy, borrow, or rent them. You can acquire some equipment as you go — renting a press at the time of fermentation or picking up some fix-it agent only when a problem arises. But you can stockpile supplies that you know you'll need so you're ready ahead of time:

✔ A container to ferment the fruit or juice in

✔ A vessel to age the wine in

✔ Tubes and hoses for moving liquid around

✔ Basic chemicals

✔ A scale

✔ Buckets (lots of buckets)

Choosing grapes

With winemaking, the ingredients are simple: grapes. Wine is fermented grape juice after all, so knowing your grapes is crucial to a satisfying finished product. The quality of a wine depends fundamentally on the quality of the grapes that go into it. Winemaking wizardry can salvage inferior grapes and make a drinkable wine, but no technology or chemical hocus-pocus can morph mediocre grapes into stunning wine.

Selecting the grapes you work with is the single most important decision you make as a home winemaker. After you settle on the source of your fruit and the grapes arrive, much of what you do as a winemaker is try to avoid screwing up the grapes. Your job is to capture all the grapes' potential and make sure stray microbes and other unforeseen factors don't spoil them.

When you become the winemaker, you gain some control over your wine supply. So think carefully about what kind of wine you like and — even more important — what you like about it. Do you tend to like wines with lots of bright acidity or not so much? Are you in love with the exotic, floral aromas of some white wines or does that turn you off? The more you know about your taste in wine, the more motivated you'll be to do the grunt work and the better you'll be at making all the little winemaking decisions that affect the final outcome.

Measuring grape chemistry

When the grapes are harvested, or at the latest, after you have them home and crushed and can measure the juice, you need to test for the basic parameters:

- **Level of ripeness,** measured in *degrees of Brix,* essentially the percentage of sugar by weight.

- **Total acidity (TA),** normally measured in grams of acid per liter.

- **pH,** a measure of the juice's electrical ionization, which you can think of (roughly) as the inverse of the total acidity. (That is, the higher the TA, the lower the pH, and vice versa.)

These three parameters have normal ranges and problem ranges, and the only way to find them out is to do simple testing. You may need to make adjustments to the grapes and juice before you do anything else: Add water if the sugar is ridiculously high, add acid if the pH is too high, and so on.

Destemming, crushing, and pressing

Home winemaking takes physical work. The three stages where grapes can't become wine without forceful human intervention are during the destemming, crushing, and pressing.

Practicing "safe" winemaking

Winemakers endlessly debate the merits of *reductive* winemaking — where oxygen is kept as far out of the picture as possible — and *oxidative* winemaking — where the useful role oxygen can play is maximized. Wine philosophers go round and round about *natural* winemaking — with little or no intervention — and *manipulative* or *industrial* winemaking — where technology is brought to bear at every point.

To maximize the chances that you'll produce a clean, stable, expressive wine and minimize the chances that you'll produce a microbial playground, we advise that you use commercial yeast strains, not whatever yeast happens to live in your garage. We want you to go out of your way to worry about oxygen exposure and temperature control and to use sulfur dioxide to safeguard your wine against alien microbes. Equally important is that you do a lot of testing to supplement what your nose and palate tell you, and, above all, remember that you can never do enough cleaning. You can make wine at home in many ways, but safe winemaking is absolutely the way to start.

You want to use the grapes, not the stems, so step one is *destemming* — stripping off the berries and separating them from the stems. Most often, the same continuous process also *crushes* the grape berries, which isn't nearly as violent as it sounds. What really happens is that the skins get cracked and the juice starts oozing out, which is a good thing because the liquid is what ferments. Gentler crushing is generally better; you're not making a smoothie.

After destemming is where white wines and reds part company. Whites generally go immediately to *pressing,* squeezing out the juice, which is clarified by gravity settling and then fermented on its own. Red wines ferment with the grape skins, allowing for extraction of color, tannin, and other goodies that won't come out on their own, no matter how nicely you ask. Reds go to press after fermentation, not before.

Juice that's ready for fermentation is known as *must.* White must is simply juice, whereas red must is juice with a slurry of skins, pulp, and seeds floating around in it. To go from must to wine, the juice goes through fermentation.

Understanding Wine Fermentation

Making a bottle of wine from start to finish may take six months or a year, but everything hinges on what happens in the week or two of *fermentation* — the period in which yeast activity extracts all the flavor, aroma, and texture goodies from the grapes and skins and converts the sugar to alcohol. Technically speaking, this is the *primary* or *alcoholic fermentation,* to distinguish it from the optional secondary, malolactic fermentation. Your job is to set it on a good course and then watch it happen.

Fermentation can be fast and furious; the snap, crackle, and pop of gases escaping from the fruit or juice can be downright noisy; the aroma can carry a block away. Fermenting white wines turn strange, greenish colors and grow a head of foam on top; they look like a pot of pea soup gone bad. But not to worry — that's just yeast at work. The spectacle of those tiny microbes making such a huge commotion is truly awesome.

Primary fermentation: From juice to wine

Grape juice exposed to air will start fermenting on its own because yeast cells and other critters that love a sugar high are everywhere. For controlled winemaking, standardized, commercial yeast strains do a much more predictable job.

Dried yeast works best when briefly rehydrated before it goes into the must. And most winemakers also add some packaged yeast nutrient to keep the little workhorses happy. The yeast cells multiply quickly, and within a day or two, their activity will be evident: You'll hear and see the must bubbling — releasing carbon dioxide — and smell wondrous fermentation aromas.

Done on a home scale — fermentations of a few hundred pounds or less — red wines take roughly a week to ferment, and white wines take two or three weeks. During that time, you have to pay attention to temperature — which generally means ensuring that red fermentation temperatures get up into the 80s and whites stay down under 60 degrees. Because the yeast and must may not produce the desired temperatures on their own, you may need to be creative with heating and cooling — electric blankets, tubs of ice water, swamp coolers, and so on. Monitor progress in the fermentation twice a day for temperature and for the drop in Brix, as more and more sugar turns to alcohol.

When white wines have fermented to *dryness* (all the fermentable sugars have gone to ethanol), the yeast slowly dies off from the lack of sugar and the alcohol's toxicity. The yeast then slowly falls to the bottom of the fermentation vessel, partially clarifying the new wine. When reds are dry, the wine gets pressed off the skins and put into containers for aging.

Knowing when it's done and what to do then

Just as fermentation lets you know it's happening, it lets you know when it's done, too. The foamy scum on the top of white wines dies down, and starting at the top, the wine slowly begins to clarify, letting some light through and accumulating the sediment of dead yeast at the bottom of the container. Reds stop hissing, and the cap of skins starts falling back into the wine soup, while the temperature drops. These are all clear signs that your wine is nearing *dryness* — the complete conversion of sugar to alcohol.

Hydrometer readings are an important indicator of the stage of your fermentation. When the hydrometer registers that your wine is down to zero Brix and zero potential alcohol, you are close to dry — but not there quite yet. The hydrometer measures the density of your sample as compared to the density of water, and when your wine has 10 to 15 percent alcohol in it, it's less dense than water — below zero on the hydrometer. (Don't worry; the hydrometer has markings for this.) No precise formula exists for how far below zero you have to get, but something in the negative 1½ to 2 degrees Brix range is the ballpark.

A more exact test for dryness, which you can easily do at home and should, uses Clinitest tablets. (These tablets are designed for testing the sugar level in urine, but pay no mind.) Following the instructions that come with the tablets, put a few drops of wine, a few drops of water, and a tablet into a small test tube and compare the resulting color to a color chart to see whether any sugar is still in the wine. Alternatively, have a winemaking shop do the analysis.

When whites get to dryness, let them sit a week or two so they can continue to clarify, and more dead yeast can head for the bottom. Keep the airlocks tightly sealed and filled with water on carboys to keep air out and the remaining carbon dioxide in as a protective blanket. If you have a canister of carbon dioxide, shoot a little gas into the top of the carboy to protect the wine from oxygen. After a while, you can rack the partially clarified wine off the bottom sludge into a new container for further aging.

Post fermentation: Completing the process

After the wine is dry, you have another round of decisions to make. You measure basic wine chemistry again, with emphasis on the post-fermentation pH and total acidity. You may need to make further adjustments.

This is also the time to take action on the mysterious matter of *malolactic fermentation* (the transformation of malic acid from the grapes into softer lactic acid by way of certain bacteria). Nearly all reds go through malolactic, and so do some whites, including most California chardonnay. Your choice is stylistic, but either way, you need to promote malo or stop it in its tracks; leaving it to chance means unstable wine that can explode in the bottle later on.

After you resolve the malolactic issue, give your wine a stiff dose of sulfur dioxide (SO_2), the amount based on the wine's pH, to shut down further microbial activity.

Also, at this point you implement your strategy for aging your wine — in carboys, stainless tanks, beer kegs, or barrels. Before heading to the chosen aging vessel, whites and reds both generally get a *racking* (siphoning the cleaner wine off of the sludgy mess at the bottom of a container) to get rid of the *gross lees,* which are the thickest, funkiest remains of the expired yeast.

Storing and Aging Your Wine

Home wine storage normally happens in one of two containers: glass carboys or oak barrels. Most home winemakers start by aging their white wines and small batches of red wine in carboys and later move up to barrels, at least for some reds. Either way makes excellent wine, with stylistic and budgetary differences. If your first wine is a modest experiment, a carboy or two is the way to go. For cases and cases of wine, barrels may be in your future.

Glass carboys (and their plastic cousins) are the default home winemaking containers. They have several advantages as a wine storage and aging solution when compared to barrel or stainless steel tank aging:

- Glass, unlike barrel oak, is inert, adding no flavors or aromas to your wine. If you want wood flavor, you can add oak chips.

- Inert glass can be sanitized better than barrels, which greatly reduces the survival chances of bad microbes.

- Glass provides a solid barrier against oxygen; the only air available is the tiny amount in the neck of the carboy and whatever oxygen gets introduced during racking.

- Carboys are smaller than barrels and more portable for filling, racking, and cleaning. A 5-gallon (19-liter) carboy filled with wine weighs roughly 50 pounds (23 kilograms) — not light, but easier to move around than a full barrel, which can weigh several hundred pounds or kilograms.

✔ Glass lets you see your wine age. You can watch it clarify, change color, and build up sludge at the bottom, signaling racking time.

✔ You need less than a hundred pounds (45 kilograms) of grapes to fill a carboy with wine — perfect for small, experimental batches or for wine-makers on a tight budget.

Naturally, carboys have their downsides, too:

✔ Glass can crack more easily than plastic, steel, or wood, which leads to leaks and the occasional small flood.

✔ Glass does such a good job of keeping out oxygen that it can limit a wine's development; barrel aging allows a small amount of oxygen to creep in, helping to round out and integrate wines.

✔ When you're making larger volumes of wine, the forest of carboys can be a pain (and take up immense storage space when empty).

✔ Glass carboys got more expensive when their prime source shifted from Mexico to Italy, raising production and transportation costs.

For making white wines on a larger scale — 30 gallons (114 liters) or more — a small stainless-steel tank may be a good investment. Most reds benefit from barrel aging, so when you're up to 15 gallons (57 liters) or more, you might add barrels to your shopping list. But carboys are part of every home cellar and show up in most commercial wineries, too.

Your main job during aging is to sniff the wine and taste it periodically and to think about whether it needs some minor tweaking. Also, make sure whatever containers you're using are topped up so that very little air is between the wine and the top of the container. Your sniffing sessions may also identify some things you'd just as soon not have in your wine, so you may need to take remedial action.

Dead yeast and other stray stuff in the wine continue to fall to the bottom (thanks to gravity), so periodically, you want to rack the wine, keeping the wine and losing the sludge. Depending on the wine, you do two, three, or four rackings (covered in the next section) over the course of three to twelve months. Time and rackings won't get your wine crystal-clear, but they'll get it most of the way there.

Finishing and Bottling

When your wine is ready for prime time — normally three to six months for whites and six to twelve months for reds — some final prep steps are in order.

When the first thick, visible layer of lees collects on the bottom of your carboy or barrel in a week or so, move the clean(er) wine away from the sludge by *racking* — siphoning the wine into another container and leaving the grosser lees behind. This is a little more complicated than it sounds.

Home winemakers do their racking with a racking tube and a length of plastic hose. Racking tubes come with *anti-sediment tips,* thimble-shaped plastic covers at the end of the tube that extend upward half an inch (1.5 centimeters) or so. When you insert the racking tube into the bottom of a sludgy carboy or barrel, the siphoned wine can only get into the tube over the brim of the plastic tip, not directly from the bottom of the container. So the sludge stays put because the wine is drawn from just slightly higher up.

For the hydraulics of siphoning to work, the source container has to be higher than the target container. Suck some air out of the end of the plastic hose, drawing wine up the racking tube and into the hose, and then insert the hose in the target. After the siphon starts, the wine will flow until it's all transferred.

Because the purpose of racking is to leave the sludge behind, try not to stir things up with the tip of the racking tube. In carboy racking, the best spot for the tip of the racking tube is along the bottom edge, the crease between the bottom and side of the carboy. After you siphon the wine halfway, tilt the carboy slowly and gently until the racking tube is straight-up vertical; that way, you stir the lees less and get the most wine with the least sludge.

Thanks to your old friend gravity, the sludge problem mostly solves itself. Millions of yeast cells are suspended in the wine at first, but sooner or later, they fall to the bottom of the container. This process takes anywhere from a couple of months to a couple of years, depending on how dirty the wine is to begin with, how much wine you have, and how you store it.

Tasting and Talking about Wine

The most important equipment you can apply to winemaking is your senses — smell, taste, and sight. Just as winemaking is in service to the grapes at hand, all the scientific biochemistry stuff is in service to the senses — how the wine smells and tastes as it develops.

You don't master the art of winemaking out of a book, not even the fine one you have in your hands. You develop your skills with your nose, your mouth, and your memory of sniffs and sips.

Aiming high

The final aspect of the home winemaking mind-set is aiming high. Set out to make something that's more than just technically wine. Set out to make good wine you can be proud of. Because you can.

The first time people make their own wine, they tend to worry that something will go haywire. But when they're done, and the finished product actually tastes like wine, it's a revelation!

(Not quite up there with giving birth, but at least as big a deal as learning to ride a bicycle.) The next year, with more confidence and tricks up your sleeve, you're going to make even tastier wine. By your third harvest, if not sooner, you'll be making wines that are varietally correct — your merlot really tastes like merlot, your pinot grigio like pinot grigio — and hold their own with the offerings at the local wine shop.

As you seek out good deals on equipment, you may want to upgrade your own sensory equipment — not by organ transplants, but by practice. One of the most useful things a winemaker can do is taste other wines, trying a variety of grapes, styles, regions, and vintages.

Taking a wine appreciation class should be mandatory for home winemakers. Classes in wine styles, wine regions, and wine tasting are readily available from community colleges, wine shops, and wine education centers; seek one out. Training in how to sniff out common wine faults — what oxidized wine or hydrogen sulfide smell like — is also valuable.

Most everyone can smell and taste, but few people are born with the vocabulary to describe what's in a glass. Putting words to wine is one of those activities that resembles dancing about architecture — it's another world and, at best, indirect. The more you do it and the more you say out loud what you're tasting, the easier it gets. Plus, to make wine you really like, you have to learn how to describe it.

Learning the lingo is important because your most important task as a home winemaker is to taste, taste, taste as your wine moves from grapes to bottle. Testing is useful to know what's going on in your fermenter or barrel, but the only way you really know your wine's mood and plans is through your nose and your mouth. This is, by the way, one of the great joys of home winemaking: tasting your wine all along the way, not just starting with the finished, bottled product, nice as that is.

Simple Grape Wine

Prep time: 10 min • **Ferment time:** 6 weeks • **Yield:** 1 gallon

Ingredients	*Directions*
6 cups white sugar	*1* Dissolve the sugar in the boiled water.
3 quarts boiled water	*2* Add the grapes to the sugar-water mixture.
1 quart ripe, mashed grapes	*3* Sprinkle yeast over all.
1 packet yeast	*4* Allow the mixture to sit for 24 hours and stir gently.
	5 Continue to stir the mixture once every 24 hours for a week.
	6 Add 1 quart boiled, cooled water to the mixture.
	7 Place the mixture in a container with an airlock and allow it to ferment for 6 weeks.
	8 Strain and rack the wine into a second container, lightly capped, for 72 hours.

Elderberry Wine

Prep time: 30 min • **Ferment time:** 11 months • **Yield:** 5 gallons

Ingredients	Directions
3 gallons black elderberries	*1* Clean the berries of all stems.
3 gallons water	*2* In a food-safe bucket, boil 3 gallons of water and pour it over the berries to cover them.
1 packet champagne yeast	
10 pounds cane sugar	*3* Cover the container loosely and allow the berries to cool and sit overnight.
	4 Remove 1 cup of the liquid and dissolve the yeast in it.
	5 Pour this yeast/liquid mixture back into the berries and water.
	6 Stir and cover the container.
	7 Allow the mixture to ferment for 72 hours, stirring every 4 hours.
	8 After 72 hours, place the cane sugar into a large kettle and add enough water so the sugar doesn't scorch and dissolves into a syrup. Cover and allow the syrup to cool to room temperature.
	9 Pour the sugar syrup into the berries and leave to ferment for an additional 5 days, stirring every 6 to 8 hours.

10 When fermentation starts to slow down, strain the mixture into a 5-gallon carboy.

11 Using the remaining berry mash, pour additional water over it and mash. Strain this water into the carboy with the first mixture, leaving a few inches of head space.

12 Insert an airlock and store for 8 weeks.

13 After 2 months, rack the wine into a clean carboy, insert an airlock, and ferment for an additional 9 months.

14 The wine is now ready to drink or bottle for longer aging.

Dandelion Wine

Prep time: 48 hours • **Ferment time:** 9.5 months • **Yield:** 1 gallon

Ingredients	Directions
1 gallon dandelion flowers (all green parts removed)	*1* In a large bowl, pour the boiling water over the flowers.
1 gallon boiling water	*2* Allow the flowers to cool and sit, loosely covered, for 48 hours.
4 organic oranges	
4 organic lemons	*3* Strain the liquid into a large glass jar or bowl.
4 pounds cane sugar	*4* Zest and juice the oranges and lemons.
1 packet yeast	
	5 Add the zest, juice, and cane sugar to the dandelion liquid. Stir until the sugar is dissolved.
	6 Sprinkle the yeast over the mixture.
	7 Cover loosely and ferment for 14 days. Stir the mixture four times a day during this time.
	8 Strain and rack for at least 9 months.
	9 The wine is now ready to drink or store for additional aging.

Chapter 16

Brewing Basics

· ·

In This Chapter

▶ Deciding what equipment and ingredients you need

▶ Ensuring the cleanliness of your homebrewing setup

▶ Brewing a batch step by step

▶ Priming and bottling your beer

▶ Taking homebrewing to the next level

· ·

*I*f you're intrigued by the idea of homebrewing, the first step is learning the science behind the art. In this chapter you find information about the basic ingredients of beer, the equipment you need to brew, and step-by-step instructions to walk you through the process.

When you feel comfortable with getting your homebrew just right, you can then start introducing your own imagination to the recipes. Chapter 17 lists some recipes to get you started.

Gathering Your Ingredients

Homebrewing is one of the most sublime hobbies. Like canning vegetables grown in your backyard garden or baking bread in your own kitchen, homebrewing enables you to recapture the hands-on rusticity of the old days while producing something that's an absolute delight to consume.

Four basic building blocks make beer — barley (malt), hops, yeast, and water. This section looks at these ingredients and their role in the brewing process.

Malt: Going with grain

Of the four main ingredients used to make beer, barley — really, grain in general — makes the biggest contribution. It's responsible for giving beer its underlying flavor, its sweetness, its body, its head of foam, and its *mouth feel* — the textural qualities of beer on your palate and in your throat, such as *viscosity* (thickness), carbonation, alcohol warmth, and so on. Grains also contribute the natural sugars that feed the yeast, which converts the sugars into alcohol and carbon dioxide during fermentation.

The word *malt* generally refers to the natural maltose sugars derived from certain grains (mainly barley) that eventually become beer. At the commercial brewing and advanced homebrewing levels, brewers produce beer through procedures that create and capture the malt sugars from the grain. At the beginner and intermediate levels of homebrewing, however, you use commercially produced malt syrup that eliminates the need for these procedures.

Before you can brew with barley, it must undergo a process known as *malting.* The malting process, simply put, simulates the grain's natural germination cycle. Fortunately for homebrewers (particularly novices), you can make beer much more easily, without having to deal with grains. You can buy a product called *malt extract,* which has been nothing less than a boon to the homebrewing industry. (Some professional brewers use it, too.)

Malt extract comes in two distinct forms. One is liquid, which is quite thick and viscous, and the other is dry, which is a rather sticky powder. Dry malt extract has a longer shelf life because of a lack of water, while liquid malt ages faster and turns darker.

Hops heaven

If malts represent the sugar in beer, hops surely represent the spice. As a matter of fact, you use hops in beer in much the same way that you use spices in cooking. The divine mission of hops is to accent the flavor of beer and, most important, contrast the malt's sweetness. This spiciness, however, isn't all hops have to contribute. Hops contribute five qualities to beer:

- A bitterness that offsets the cloyingly sweet flavor of malt
- A zesty, spicy flavoring that accents the malt character in beer
- A pungent floral/herbal aroma
- Bacterial inhibitors
- Natural clarifying agents

Traditionally speaking, brewers handpicked hops from the vine and air-dried them in bulk before tossing them whole into the brew kettle. Today, however, hops are processed and sold in three different forms: whole-leaf hops, pellets, and hop extract.

More than 70 recognized hop varieties exist, and each hop variety offers different nuances in bittering intensity, flavor, and aroma. The differences among them are often so subtle that even the most experienced brewers and beer judges are hard-pressed to recognize their individual attributes in a given beer — especially when you consider that brewers often use blends of different hops in a single batch of beer.

How and when you use the hops determines the effect they have on the finished brew. The longer you boil the hops, the more bitterness dissolves into the *wort* (up to a point). Boiling hops for 5 to 30 minutes imbues the beer with far less bitterness than the hops could potentially add, but you get some hop flavor. Adding hops very late in the boil and boiling them for less than 5 minutes provides the beer with aromatics and little else. Many homebrew recipes tell you not only which hop varieties to use and in what quantities but also how long to boil them to get bittering, flavoring, or aromatizing characteristics from them.

Yeast: The key to fermentation

Although yeast is an ingredient that the average beer consumer rarely contemplates, brewers often consider it the most important ingredient. In fact, yeast can have a greater influence on the finished beer than any other single ingredient.

Yeast is a member of the fungus family. It's a living, single-celled organism and one of the simplest forms of life. Because it has cell-splitting capabilities, it's also self-reproducing. Yeast is the one ingredient responsible for carrying out the fermentation process in brewing.

Fermentation, simply put, is the natural conversion of sugar to alcohol. Yeast has a voracious appetite for sweet liquids. And, in exchange for a good, sweet meal, yeast produces equal amounts of ethanol (ethyl alcohol) and carbon dioxide.

Fermentation is, indeed, magical and mystical. A simple yeast cell consumes sugar (in liquid form) and in turn excretes alcohol and carbon dioxide, in addition to hundreds of flavor compounds. As part of the growth process, a single cell reproduces by cloning itself — splitting into two separate cells. Multiply this chain of events by billions, and you have fermentation.

Yeast for the homebrewer comes in both a dry form and liquid form. Because of its convenience, we highly recommend dry yeast at the beginner level. Dry yeast comes in granular form in small foil packets. You simply tear these packets open and rehydrate the yeast in water before pitching it into fresh wort. However, because of dry yeast's lack of stylistic variety, you want to progress to liquid yeast cultures as soon as you're comfortable.

Don't forget the water

Water is just one of the four primary ingredients used to make beer, but considering that it constitutes up to 95 percent of a beer's total ingredient profile, water can certainly have a tremendous influence on the finished product. The various minerals and salts found in water used for brewing can accentuate beer flavors or contribute undesirable flavor components.

Having said that, however, you can still make good beer with average tap water. Thousands of homebrewers are proving it every day. A general rule says, "If your water tastes good, so will your beer." A caveat is important here, though: This general rule pertains solely to extract-based homebrews.

Certain aspects of water composition become much more important when homebrewers begin mashing their own grains. And water chemical profiles are really important to the small percentage of homebrewers who are determined to imitate the water found in famous brewing cities around the world.

If your water is from a private underground well, it may be high in iron and other minerals that may affect your beer's taste. If your water is softened, it may be high in sodium. If your water is supplied by a municipal water department, it may have a high chlorine content. Other than chlorine, the filtering (the primary method of removing elements and impurities from water) performed at municipal water sources usually produces water that's sufficiently pure for brewing.

High iron, sodium, and chlorine contents in your brewing water aren't desirable. If these minerals are present in your brewing water, you may want to consider buying bottled water for your brewing needs.

Cleaning Up Your Act: Sanitation

Scrupulously clean brewing equipment and a pristine brewing environment are the keys to making good beer. And by clean, we don't mean just soap-and-water clean. In homebrewing, serious sanitation is necessary.

Take a closer look at a few important words used in this chapter:

- ✔ **Clean:** As it pertains to homebrewing, clean means that you've removed all dust, dirt, scum, stains, and other visible contaminants from your brewing equipment and bottles to the best of your ability.

- ✔ **Sanitize:** After the visible contaminants are history, you need to sanitize your equipment and bottles. Sanitation is the elimination of invisible contaminants (bacteria and other microorganisms) that can ruin your brew. Clean requires a little elbow grease; sanitized requires chemicals.

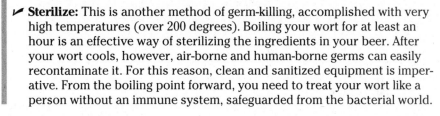

  Never assume that clean equipment is sanitized, and never assume sanitized equipment is clean. For the good of your beer, practice good cleaning and sanitizing techniques — in that order.

- ✔ **Sterilize:** This is another method of germ-killing, accomplished with very high temperatures (over 200 degrees). Boiling your wort for at least an hour is an effective way of sterilizing the ingredients in your beer. After your wort cools, however, air-borne and human-borne germs can easily recontaminate it. For this reason, clean and sanitized equipment is imperative. From the boiling point forward, you need to treat your wort like a person without an immune system, safeguarded from the bacterial world.

Nothing is more important to your production of a clean, drinkable, and enjoyable beer than utilizing proper sanitizing procedures prior to brewing. You must sanitize or sterilize anything and everything that will come into contact with your beer at any time.

"What's the big deal?" you may ask. Well, millions of hungry microbes just love to make meals of freshly brewed beer. These microbes are in your home, on your body, and even in the air you breathe (cough, wheeze). Bacteria and fungi are the forms of microbes that you need to be wary of. They're both opportunistic, and if you let them have their way with your brew, they will (always with negative results). Beer that bacteria have contaminated smells and tastes awful.

So, how can you get rid of these little beer-ruining pests? Well, in truth, you can't really get rid of them completely; the idea is to keep them away from your beer (or at least minimize their effect). Here are some helpful tips to assist you in deterring these foes:

- ✔ Keep your brewery (kitchen, laundry room, basement, or wherever you make your beer) as clean and dust-free as possible.

- ✔ Quarantine all furry, four-legged family pets in another part of the house while you brew or bottle your beer.

- ✔ Consider every cough and sneeze a threat to your beer.

✔ Treat your equipment well. Clean and sanitize it properly prior to brewing, rinse it well, and dry it off after every use and before storing it away. Keep your equipment stored in a dust- and mildew-free location if at all possible. You may even want to go as far as sealing all your equipment in large-capacity garbage bags between brewing sessions.

Chemicals that you can use to clean and sanitize your homebrewing equipment include iodine-based products, chlorine-based products, caustics (which can burn your skin), ammonia, and a couple of environmentally safe cleansers that contain oxygen-based percarbonates. For our money, ordinary, unscented household bleach is still the best bet.

Practicing safe sanitation

The most effective methods of sanitizing involve soaking rather than intensive scrubbing. For this purpose, the best place to handle sanitizing procedures is in a utility basin or large-capacity sink.

Never use any abrasives or materials that can scratch your plastic or metal equipment, because pits and scratches are excellent hiding places for those wily bacteria. Using a very soft sponge that you devote solely to cleaning homebrew equipment is a good idea.

When you're ready to begin sanitizing, follow these instructions:

1. **Place the items you want to sanitize in the plugged utility basin and begin drawing cold water into the basin.**

   Because chlorine is volatile, don't use hot water with bleach; the heat of the water causes the chlorine gas to leave the water much more quickly. If you're sanitizing the fermenter, carboy, or bottling bucket, you need to fill only those items rather than the whole basin or sink. You can place smaller items in the sanitizing solution within these larger items. However, for bottles, you need to fill the entire sink.

2. **As the water runs, add cleansing/sanitizing chemicals according to package directions or pour in 1 ounce of unscented household bleach per gallon of water.**

3. **Completely immerse all the items you want to sanitize in the sanitizing solution.**

   Don't forget to include the fermenter lid, which you have to force into the fermenter sideways. Always allow 30 minutes for all bottles and equipment to soak.

4. **After 30 minutes, remove your equipment and thoroughly rinse the various pieces in hot water.**

5. **Sanitize the spigots on the fermenter and bottling bucket by draining the sanitizing solution through each spigot.**

6. **Allow everything to air dry.**

   The fermenter lid, placed upside down on a clean surface, is a good place to dry the smaller sanitized items.

Bottle cleanliness is a virtue

When you're ready for bottling, you need to clean and sanitize your bottles. You can clean your bottles pretty much at your convenience if you store them properly, but sanitizing your bottles too far ahead of time may lead to bacterial recontamination. Read more on sanitizing bottles in the section "Bottling Your Brew" later in this chapter.

Ready, Set, Brew: Beginners

Beginning brewers have to start somewhere, and that somewhere needs to be with an all-inclusive homebrewing kit. A kit is simply a package you buy from a homebrew supply store that includes all the ingredients (pre-hopped malt extract and a packet of yeast) that you need to brew a particular style of beer. The kits are ingredient kits only, mind you; they don't include any equipment.

But brewing beer from a kit also has a possible downside, depending on your perspective. At the beginner homebrewing level, you have little personal control over most of the beer-making process. When you use a kit, much of the thinking and the work have been done for you. This extra guidance is good if you're just starting out in the world of homebrewing and want to work with a net the first few times through the process. You can master several aspects of homebrewing (sanitation, racking, observing fermentation, and bottling) at the beginner level. Thus, another simple rule at the beginner level: The more you brew, the better your beer gets. It's a delicious circle.

Assembling your tools

Before you start the brewing process, make sure you have all your homebrewing equipment, you've properly sanitized it (see the earlier "Cleaning Up Your Act: Sanitation" section), and it's in place and ready to use. Here's a quick equipment checklist for you:

- Airlock
- Brewpot with lid

    The size of your brewpot depends on the amount of brew you're making. Because you also boil the brew in this pot, buy a larger one than you think you'll need. For instance, for a 5-gallon recipe, use at least an 8-gallon brewpot.

- Brew spoon
- Coffee cup or small bowl (for proofing the yeast)
- Fermentation bucket with lid
- Hydrometer (the hydrometer cylinder isn't necessary at this time)
- Rubber stopper (to attach to the airlock)

Now you just need a couple of simple household items to complete the ensemble. Gather and sanitize a long-handled spoon or rubber spatula (for scraping the gooey malt extract from the cans), hot pads (to hold on to hot pots and pans), and a small saucepan (to heat up the cans of extract). Speaking of which, be sure to have your homebrew kits on hand.

We recommend either two 3.3-pound cans of pre-hopped malt extract (plus yeast) or one 6.6-pound can. This is the appropriate amount for the average 5-gallon batch of beer. The style of beer or brand of malt extract is your choice.

Brewing your first batch

The following numbered list covers 24 steps that walk you all the way through the brewing process. Twenty-four steps may sound pretty intense, but they're easy, quick, and painless steps.

1. **Fill your brewpot about two-thirds full with clean tap water or bottled water, and then place it on the largest burner of your stove.**

    The exact volume of water isn't terribly important during this step because you add cold water to the fermenter later to bring the total to 5 gallons.

2. **Set the burner on medium-high.**

3. **Remove the plastic lids from the kits and set the yeast packets aside.**

4. **Strip the paper labels off the two cans of extract and place the cans in a smaller pot or saucepan filled halfway with tap water.**

    Place the pot or saucepan on another burner near the brewpot. The water's purity isn't important here because you don't use this water in the beer.

5. **Set the second burner on medium.**

6. **Using hot pads, flip the cans in the warming water every couple of minutes.**

7. **As the water in the brewpot begins to boil, turn off the burner under the smaller pot (containing the cans), remove the cans from the water, and remove the lids from the cans.**

8. **Using a long-handled spoon or rubber spatula, scrape as much of the warmed extract as possible from the cans into the water in the brewpot.**

9. **Immediately stir the extract/water solution and continue to stir until the extract completely dissolves in the water.**

   This malt extract/water mixture is now officially called *wort*. Stir the wort immediately or you risk scorching the extract on the bottom of the brewpot.

10. **Top off the brewpot with more clean tap or bottled water, keeping your water level at a reasonable distance — about 2 inches — from the top of the pot to avoid boilovers.**

11. **Bring the wort to a boil.**

    Turn up the burner, if necessary.

12. **Boil the wort for about an hour, stirring the pot every couple of minutes to avoid scorching and boilovers.**

13. **Turn off the burner and place the lid on the brewpot.**

14. **Put a stopper in the nearest sink drain, put the covered brewpot in the sink, and fill the sink with very cold water.**

    Fill the sink completely (or to the liquid level in the brewpot if the sink is deeper than the brewpot).

15. **After 5 minutes, drain the sink and refill it with very cold water.**

    Repeat as many times as you need until the brewpot is cool to the touch.

16. **While the brewpot is cooling in the sink, draw at least 6 ounces of lukewarm tap water into a sanitized cup or bowl.**

17. **Open the yeast packets and pour the dried yeast into the cup or bowl of water.**

    This process, called *proofing,* is a gentle but effective way to wake up the dormant yeast and ready it for the fermentation to follow.

18. **When the brewpot is relatively cool to the touch, remove the brewpot lid and carefully pour the wort into the fermentation bucket.**

    Make sure the spigot is closed.

19. **Top off the fermenter to the 5-gallon mark with cold, clean water, pouring it vigorously into the bucket.**

   This splashing not only mixes the wort with the additional water but also aerates the wort well. The yeast needs oxygen to get off to a good, healthy start in the fermentation phase. Because boiled water is virtually devoid of oxygen, you need to put some oxygen back in by aerating the wort.

   Failure to aerate may result in sluggish and sometimes incomplete fermentations.

20. **Take a hydrometer reading.**

   See the section "Brewing-day reading," later in this chapter, for specific information about this process.

21. **After you take the hydrometer reading and remove the hydrometer, pour the hydrated yeast from the cup or bowl into the fermenter and give it a brisk stir with your brew spoon.**

22. **Cover the fermenter with its lid and thoroughly seal it.**

23. **Put the fermenter in a cool, dark location, such as a basement, a crawl space, or an interior closet.**

24. **After the fermenter is in a good place, fill the airlock halfway with water and replace the cap, and attach the rubber stopper and position it snugly in the fermenter lid.**

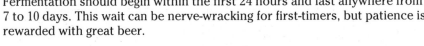

   Check to make sure that the fermenter and airlock are sealed airtight by pushing down gently on the fermenter lid. This gentle pressure causes the float piece in the airlock to rise; if it doesn't, you have a breach in the seal. Recheck the lid and airlock for leaks.

Fermentation should begin within the first 24 hours and last anywhere from 7 to 10 days. This wait can be nerve-wracking for first-timers, but patience is rewarded with great beer.

How quickly the beer begins to ferment and how long the fermentation lasts depends on the amount of yeast, the yeast's health, the temperature at which the beer is fermenting, and whether the wort was properly aerated. Healthy yeast, mild temperatures (65 to 70 degrees), and an abundance of oxygen in the wort make for a good, quick ferment. Old, dormant yeast, cold temperatures, and under-oxygenated wort cause fermentations to start slowly, go on interminably, or even quit altogether.

After your beer has been in the fermenter for about a week or so, check the bubbling action in the airlock. If visible fermentation is still taking place (as evidenced by the escaping bubbles), continue to check the bubbling on a daily basis. When the float piece within the airlock appears to be still, and the time between bubbles is a minute or more, your beer is ready for bottling. Before you begin the bottling procedures, however, you need to take a second gravity reading to make sure that the fermentation is complete. (See the section "Pre-bottling reading" later in this chapter for specifics.)

Taking hydrometer readings

This section shows you specifically how to take hydrometer readings on brewing and bottling days.

Brewing-day reading

You want to take the first hydrometer reading on brewing day (see Step 20 in the earlier "Brewing your first batch" section of this chapter). To take a good reading, do the following:

1. **Lower the sanitized hydrometer directly into the cooled and diluted wort inside the fermenter.**

   As you lower the hydrometer into the wort in the fermenter, give it a quick spin with your thumb and index finger; this movement dislodges any bubbles clinging to the hydrometer that may cause you to get an incorrect reading.

2. **Record the numbers at the liquid surface on the hydrometer scales.**

   You need this information on bottling day to decide whether fermentation is complete and to figure out the alcohol content in your beer. The gravity of your malt extract and water mixture determines the numbers you see on the scales. Typically, 6 or so pounds of malt extract diluted in 5 gallons of water appear on the hydrometer's O.G. scale as 1.048, and the alcohol potential (as noted on the hydrometer's alcohol potential scale) is around 6 percent.

Pre-bottling reading

When you think your beer is ready for bottling (based on the bubbling action in the airlock), it's time to take another hydrometer reading. When compared to the first (brewing day) reading, this reading helps you decide whether your beer is actually ready to be bottled.

1. **With your hydrometer test cylinder in hand, take a sample of beer from the fermenter's spigot.**

   Fill the cylinder to within 1 inch of the opening, leaving room for liquid displacement of the immersed hydrometer.

Calculating alcohol content percentage

Here's a way to help you figure out how much alcohol has been produced in your beer during fermentation. If you use about 6 pounds of liquid malt extract for your brew, your original gravity (or *O.G.* in homebrew lingo) is in the neighborhood of 1.048. If your yeast is good and hungry, your final gravity *(F.G.)* a week or so later (after fermentation) will be about 1.012. Because a *ten-forty-eight* (also acceptable homebrewer lingo for the 1.048 gravity) represents an alcohol potential of 6 percent, and a *ten-twelve* (the 1.012 gravity) represents an alcohol potential of 2 percent, the yeast produced 4 percent alcohol in your brew. (Subtract the final alcohol potential from the original alcohol potential to derive the alcohol content percentage.)

2. **Immerse the hydrometer in the beer, record the numbers at the liquid surface of the hydrometer, and compare the numbers with the numbers you recorded on brewing day.**

Remember that the average healthy yeast consumes at least 65 percent of the available sugars in the wort. If the final gravity reading on your fermented beer isn't 35 percent or less of the original gravity, too much natural sugar may be left in your beer.

Here's a sample equation: If your beer has an original gravity of 1.048, subtract 1 so that you have 0.048; then multiply 0.048 by 0.35, which results in 0.017. Now add the 1 back in. If the final gravity of your beer is higher than 1.017, you want to delay bottling a few more days.

After you take your hydrometer reading, *don't* pour the beer from the cylinder back in with the rest of the beer. If you do so, you risk contaminating your beer.

A Primer on Priming

As you may know, yeast is responsible for producing the natural carbonation in beer and other fermentable beverages. And because carbon dioxide escapes from the fermenter, you may be wondering how you can put the carbonation back into your beer. The answer is a simple little trick called priming.

Priming means adding more fermentable sugar to the beer just prior to bottling it. The small number of yeast cells that remain in the solution when you bottle your brew eat the sugars you add and thus provide the desired carbonation within the bottle. Of course, this process doesn't happen overnight. You need to allow another one to two weeks before your beer is properly carbonated and ready to drink.

Getting ready to prime

When you choose a priming sugar for beer, you need to consider two things: the quantity and the fermentability of the priming ingredient you're using. The first factor is dependent on the second factor.

The idea of priming is to put more fermentable sugar into the beer so that the yeast can create the needed carbonation within the bottle. But consider the consequences of giving the yeast too little or too much sugar to eat: Too little priming sugar can result in an under-carbonated beer, but too much sugar may cause your bottles to explode! Don't panic — we cover proper priming quantities in the next section.

Two important points regarding beer priming: First, not all beer styles contain the same level of carbonation, and second, high-gravity beers you intend to condition in the bottle over longer periods of time are likely to build up increasing carbonation levels as they age.

Making primer decisions

Different sugar sources (refined sugar, honey, molasses, and so on) have different levels of fermentability, which means that the more fermentable sugar a priming mixture has, the less of the mixture you need. Conversely, you need more of a priming mixture with less fermentable sugar. Note that because you use them in such small quantities, priming sugars have virtually no effect on the beer's taste.

To keep the priming process simple, use dextrose (corn sugar). Dextrose is highly fermentable, widely available, easy to work with, and inexpensive.

Regardless of what form of priming sugar you use to prime your beer, always dissolve and dilute it in boiling water first and allow the priming mixture to cool before adding it to the beer. The amount of water you use to boil the sugar is of little concern — 1 or 2 cups is typical.

In the average 5-gallon batch of beer, ¾ of a cup (*not* ¾ of a pound!) of dextrose is the maximum recommended amount for priming. Using more than this may result in exploding bottles but is more likely to result in over-carbonated beer and *gushers* — bottles that act like miniature volcanoes when your buddies pry the caps off. Hmmm, sounds like a good prank, come to think of it.

Bottling Your Brew

Bottling homebrew isn't a difficult procedure, but brewers often deride it as tedious at worst and boring at best. But for millions of people who brew their beer at home, bottling represents the only option for packaging their finished brew.

Picking out bottles

Using good, sound, safe bottles is an important part of the bottling process. Your beer may start out as a world-class brew, but if the bottles aren't worthy, you can end up with leaks, explosions, and sticky messes. When you're buying bottles, don't skimp.

Bottles come in all sizes and shapes, but check out these important suggestions. Your homebrew bottles

- ✔ **Should be the thick, returnable type (no cheap throwaways).** The thick, returnable-type bottles can withstand repeated uses; cheap throwaways mean thin glass and easy breakage.

- ✔ **Should be made of colored glass (the darker the better).** Light damages beer; tinted glass protects against light damage.

- ✔ **Shouldn't have a twist-off opening.** Bottle caps can't seal across the threads on twist-off bottles.

- ✔ **Should be of uniform size.** Although uniform bottles aren't a requirement, having all your beer in bottles of the same size and shape makes capping and storing much easier.

Using larger bottles is a way to expedite the bottling process and free you from its drudgery. The more beer the bottles can hold, the fewer bottles you need.

Preparing to bottle

Before you start any bottling procedures, take a hydrometer reading of the beer in the fermenter to verify that fermentation is sufficiently complete. Just steal a little beer out of the spigot to fill the hydrometer cylinder to within an inch of the top (but no more). (See the earlier section, "Pre-bottling reading," for more information.)

After you've made certain that the beer is done fermenting, retrieved the bottling equipment, and quarantined the family pets, you're ready to start the bottling process. As always, setup starts with sanitizing all the necessary equipment (see the "Cleaning Up Your Act: Sanitation" section earlier in this chapter). Here's the equipment you need:

- A bottling bucket
- A racking cane (if bottling from a vessel without an attached spigot)
- A plastic hose
- A bottling tube (with spring-action or gravity-pressure valve)
- Bottles (enough to hold 640 ounces of beer)
- Bottle caps (enough to cap all your bottles, plus some extras — just in case)

You need to sanitize all these items before bottling, so you also need a sanitizing agent. You also need the following:

- A bottle brush
- A bottle rinser
- A bottle capper
- Two small saucepans
- ¾ cup dextrose (corn sugar) for priming (see the section "A Primer on Priming" earlier in this chapter)

Now, here are the steps for the bottling brigade:

1. **Fill your utility tub or other designated sanitizing basin with enough cold water to cover your submerged bottles, adding bleach or another sanitizing agent according to the package directions.**

2. **Submerge as many bottles as you need to contain your full batch of 5 gallons of beer.**

   Make sure that your bottles are scum-free before dunking them in the sanitizing solution. You need to separately scrub any bottle with dried or living crud in the bottom with a cleanser such as trisodium phosphate (TSP) before you sanitize it.

   You can fill and submerge the bottles in less than half the time if you place a drinking straw in the bottles; the straw enables the air within the bottle to escape through the straw instead of slowly bubbling through the opening (your bottling tube with the valve detached suffices here).

3. **Allow your bottles to soak for at least half an hour (or the time necessary according to package directions).**

4. **While the bottles soak, dissolve ¾ cup of dextrose in a pint or so of water in one of the saucepans, cover the solution, and place it on a burner over low heat.**

5. **Put your bottle caps into your other saucepan, fill the pan with enough water to cover all the caps, and place the pan on another burner over low heat.**

Put enough bottle caps for as many bottles as you have soaking, plus a few extra; having too many sterilized caps ready for bottling is better than not having enough.

6. **Allow both pans to come to a boil, remove them from the heat, and allow them to cool.**

7. **After the bottles soak for half an hour, connect the bottle rinser to the faucet over the sanitizing tub.**

8. **With one hand over the opening (so that you don't get squirted), turn on the hot water.**

After the initial spray, the bottle washer holds back the water pressure until a bottle is lowered over the stem and pushed down.

9. **Start cleaning the bottles one by one with the bottle brush and then drain the sanitizer, rinse your bottles with the bottle rinser, and allow them to air dry.**

Continue this step until all bottles are clean. Visually check each bottle for cleanliness rather than just assume that they're all clean.

Four dozen free-standing bottles make one heck of a breakable domino effect. Always put your cleaned bottles back into six-pack holders or cardboard cases to avoid an aggravating and easily avoidable accident.

10. **Drain the utility tub of the bottle-cleaning water.**

11. **Place the bottling bucket in the tub and fill it with water and the sanitizing agent of your choice.**

12. **Place the bottling hose, bottling tube, and hydrometer cylinder into the bottling bucket and allow them to soak for half an hour (or according to sanitizing agent directions).**

13. **While the bottling equipment soaks, retrieve the still-covered fermenter from its resting place and place it on a sturdy table, counter, or work surface about 3 or 4 feet off the ground.**

At this point, you need to set up your bottling station, making sure that you have the priming sugar mixture (still in the saucepan), bottle caps, bottle capper, bottles, and hydrometer with cylinder on hand.

If you're bottling your brew directly from the primary fermenter, you want to have already taken a hydrometer reading to confirm completion of fermentation. If you're bottling from your secondary fermenter (glass carboy; see Chapter 7), incomplete fermentation isn't a concern, and you can take a hydrometer reading (to determine final gravity and alcohol content) as the beer drains into the bottling bucket.

14. **After half an hour, drain the sanitizing solution from the bottling bucket through the spigot on the bottom and, after the bucket is empty, thoroughly rinse the remaining pieces of equipment (hose, bottling tube), along with the bottles and caps, and bring them to your bottling station.**

Pouring a cold one: Getting your beer into bottles

As soon as your bottling station is all set up and ready to go, start bottling your brew by following these steps:

1. **Place the bottling bucket on the floor directly below the fermenter and connect the plastic hosing to the spigot on the fermenter, allowing the other end of the hosing to hang inside the bottling bucket.**

   If you're initiating the bottling procedures from your glass carboy (see Chapter 7), you can't rely on the convenience of a spigot to drain out the beer. You need to use your racking cane and siphon the brew.

2. **Pour the dextrose and water mixture into the bottling bucket.**

   See the earlier section, "A Primer on Priming," for more on dextrose. The dissolved corn sugar mixes with the beer as the beer drains from the fermenter into the bottling bucket. After you bottle the beer, this sugar becomes another source of food for the few yeast cells still remaining in the liquid. As the yeast consumes the sugar, it produces the beer's carbonation within the bottle. Eventually, the yeast again falls dormant and creates a thin layer of sediment on the bottom of each bottle.

   If, by chance, you bottle a batch of beer that isn't fully fermented or you somehow add too much dextrose at bottling time, you may find out firsthand what a mess exploding bottles can make. Excess sugar (whether added corn sugar or leftover maltose from an unfinished fermentation) overfeeds the yeast in an enclosed bottle. With nowhere for the pressure to go, the glass gives before the bottle cap. Kaboom! Mess! Do not overprime. (Use no more than ¾ cup of dextrose in 5 gallons of beer.) See the "A Primer on Priming" section earlier in this chapter for more advice on priming.

3. **Open the spigot on the fermenter and allow all the beer to run into the bottling bucket.**

   Don't try to salvage every last drop from the fermenter by tilting it as the beer drains down the spigot. The spigot is purposely positioned about an inch above the bottom of the fermenter so that all the spent yeast and miscellaneous fallout remains behind.

4. **After the last of the beer drains, close the spigot, remove the hose, and rinse it.**

   Avoid splashing or aerating your beer as you bottle it. Any oxidation that the beer picks up now can be tasted later. Yuck.

5. **Carefully place the bottling bucket up where the fermenter was, connect the rinsed hose to the spigot on the bottling bucket, and attach the bottling tube to the other end of the hose.**

6. **Arrange all your bottles on the floor directly below the bottling bucket.**

   Keep them all in cardboard carriers or cases to avoid potential breakage and spillage.

7. **Open the spigot on the bottling bucket and begin to fill all the bottles.**

   Gently push the bottling tube down on the bottom of each bottle to start the flow of beer. The bottle may take a short while to fill, but the process always seems to accelerate as the beer nears the top. Usually, a bit of foam rushes to the top of the bottle; don't worry! As soon as you withdraw the bottling tube, the liquid level in the bottle falls.

8. **Remove the tube from each bottle after foam or liquid reaches the top of the bottle.**

   After you remove the bottling tube from the bottle, the beer level falls to about an inch or so below the opening. Homebrewers have differing opinions on how much airspace (or *ullage*) is necessary. Some say the smaller the airspace, the less oxidation that can occur. Others claim that if you don't have correct ullage, the beer can't carbonate properly. Rather than jump into the fray, we say that if it looks like the space in bottles of beer from commercial breweries, go with it!

9. **After you completely drain the bottling bucket, close the spigot, remove the hose, toss it inside the bottling bucket, and set everything aside to be cleaned after you complete all the bottling procedures.**

10. **Place all the bottles on your tabletop or work surface; place a cap on each bottle, position your bottles in the capper (one at a time), and pull down on the capper handle or levers slowly and evenly.**

    You may want to do this task as soon as each bottle is full as insurance against everything that can go wrong when full bottles of your precious brew are sitting around open.

Both bench- or two-handle-style cappers come with small magnets in the capper head designed to hold and align the cap as you start crimping. We don't trust the magnet to hold the caps in alignment and prefer to seat them on the bottles by hand.

Occasionally, a cap may crimp incorrectly. If you suspect that a cap didn't seal right, tilt the bottle sideways and check for leakage. If you find you have a leaker, yank the cap and replace it. (You boiled extras — right?)

11. **Your homebrew needs to undergo a two-week conditioning phase, so store your liquid lucre in a cool, dark location (such as the same place that you kept the fermenter).**

This phase is where the remaining yeast cells chow down on the dextrose and carbonate your beer.

Putting your brew in the fridge isn't a good idea — at least for the first two weeks — because the very cold temperatures stunt the yeast's carbonating activity.

12. **Thoroughly rinse your brewing equipment in hot water and store it in a place that's relatively dust- and mildew-free; you may even want to go that extra step and seal all your equipment inside a large-capacity garbage bag.**

This step may be the most important one of all, not so much for the brew just made but for the next one. Consider this step an insurance policy on your next batch of beer. Like most insurance policies, it's boring but worthwhile.

After two weeks pass, check to see whether the bottles have *clarified* (the yeasty cloudiness has settled out). Chill a bottle or two for taste-testing. Like any commercial beer, you need to decant homebrew before drinking, not only to release the carbonation and the beer's aromatics but also to pour a clear beer. Drinking homebrew out of the bottle stirs up the sediment, creating a hazy beer.

Intermediate Homebrewing

The basic differences between beginner and intermediate homebrewing can be easily — but not completely — summed up in a few lines. Intermediate brewers can

- Use specialty grains to add more color and flavor in the beer
- Choose and add hops as opposed to using hopped extract

    ✔ Use liquid yeast cultures instead of freeze-dried yeast

    ✔ Perform secondary fermentation procedures

The combination of ingredient changes and new procedures can make a world of difference in the quality of your homebrew. And you don't even have to apply these changes all at once.

Using better ingredients

Your first step away from being a novice brewer is to take effective but simple measures toward improving your beer. The first of these measures has to do with adding more and better ingredients.

Whenever you buy homebrewing ingredients, make sure you store them properly if you aren't going to use them immediately. You need to refrigerate all grains, hops, and yeast packets. (Freeze the hops if you plan to store them long term.) Never allow any of your ingredients to lie around in a warm environment or in direct sunlight, even if they ask you nicely. Think of beer ingredients as food products, and think of the way most food products decay over time — especially in warmer environments.

Conditioning for better beer with secondary fermentation

Using different ingredients is only one way that intermediate homebrewers set themselves apart from the beginners. The other way is by using different methods of conditioning.

Conditioning means allowing your beer additional time to mature, mellow, clarify, and carbonate.

Secondary or *two-stage fermentation* is all about conditioning your beer. When you brew at the beginner level, you put the fresh wort in the primary fermenter, let the yeast do its thing, and then bottle the beer. The beer has about two weeks to condition in the bottle before you start drinking it. You did the right thing (within the limitations of your equipment and expertise), but now you can do more.

At the beginner level, taking the freshly fermented beer out of the primary fermenter was necessary not just because the initial fermentation was over, but also because all those little yeasties, fresh from a gluttonous feast, were about to start decomposing. That's right, enzymes in the sugar-starved yeast begin to break down the yeast cells. This horrific event is called yeast *autolysis.* Autolysis can impart a sulfury, rubbery stench and flavor to your beer. So leaving your fresh, young beer sitting on that bulging layer of self-destructing yeast dregs is akin to allowing your child to wallow with pigs in the mud — and you don't want to smell either one of them when they're done. Racking your beer over to a secondary fermentation vessel effectively leaves most of the sedimented yeast and other organic matter behind.

So if bottling the beer after one week worked before, why can't it now? It still can, but now that you're introducing more ingredients into the brewpot, the added flavors and textures in your beer need more time to blend together. By allowing the beer to undergo a secondary fermentation, you promote a mellowing process that makes a noticeable improvement in your beer.

Chapter 17

Brewing Beer

This chapter encourages you, the budding homebrewer, to get more personally involved in the brewing process. You discover information on what makes a beer an ale, a lager, or something else entirely. This chapter also includes a variety of recipes to get you going on your brewing journey.

Looking at Beer Types

Before you join in on the fun of homebrewing, you should have a basic understanding of the cast of characters: the types of beers.

Ales

Ales, by traditional definition, are beers fermented with top-fermenting yeast (it rises to the surface during fermentation) at warm temperatures for relatively short periods of time. Ales are primarily associated with England and Ireland, but you can find them in a wide variety of styles in most brewing nations, such as Australia, Belgium, Canada, and the United States. Some examples include

‣ American ale

‣ Belgian ale

- Brown ale
- Extra special/strong bitter (ESB)
- India pale ale
- Irish red ale
- Porter
- Stout

The recipes in this chapter cover some of these ales. Because certain beer styles are very difficult to produce at the homebrewing level, we emphasize popular beer styles.

Lagers

Lagers are beers fermented with bottom-fermenting yeast (it settles at the bottom during fermentation) at cool temperatures for relatively long periods of time. They're primarily associated with northern and eastern European countries, such as the Czech Republic, Denmark, Germany, and the Netherlands. In fact, they're produced in a wide variety of styles in most brewing nations, including Australia, Belgium, Canada, China, France, Ireland, Japan, Mexico, New Zealand, Russia, the United Kingdom, and the United States. Examples include the following:

- American lager
- Bock beers
- Bohemian pilsner
- German dark lager
- Märzen/Oktoberfest

This chapter provides recipes for three lager varieties that you can produce at the homebrewing level.

Mixed beers

Mixed-style beers, or *hybrids,* are beers that cross the lines between conventional beer styles. These beers are fermented and aged with mixed traditions; one beer may be fermented with ale yeast at cold temperatures, and another fermented with lager yeast at warm temperatures. Sometimes — depending on the brewer's whims — they may be fermented either coolly or warmly as either ales or lagers. Some examples:

- ✔ Christmas and winter seasonal beer
- ✔ Fruit and vegetable beer
- ✔ Smoked beer
- ✔ Wheat beer

This chapter provides two recipes for beer styles that don't fit neatly into the regular ale and lager categories.

Exploring Specific Types of Ales, Lagers, and More

After you have the basics of homebrewing in place, it's time to widen your palate with an endless variety of brews. Travel the world of ales, lagers, and one-of-a-kind herb and spice beer. Enjoy re-creating tastes from other countries, rich in both flavor and history.

Irish red ale

Traditional Irish ales are easy-drinking but full-flavored, malt-accented brews. Consider the following:

- ✔ **Appearance and flavor profile:** Amber to deep copper/red in color (thus the name). This style's caramel malt flavor and sweetness sometimes has a buttered-toast or toffee quality. Irish ales often exhibit a light taste of roasted malt, which lends a characteristic dryness to the finish.

- ✔ **Commercial examples:** Smithwick's Irish Ale, Ireland; Kilgubbin Red Ale, Chicago.

If you're a beginner brewer and want to make an Irish red ale, you can use either pale extract or amber hopped extract; both finish within the generous color range for this style. If you choose the pale extract, try getting a little kettle caramelization by vigorously boiling the wort an extra 30 minutes or more. When you get a little more experience, try making the Irish red ale recipe later in this chapter.

American pale ale

Americans have been brewing ale since the first wave of colonists reached the shores of the New World in the 1600s. With the recent upsurge in small-batch brewing, American ales are now leading the microbrewing renaissance.

- ✔ **Appearance and flavor profile:** Pale to deep amber to copper color. These beers are medium-bodied, offer medium hop flavor and aroma, and are fruity and estery. Expect low to medium maltiness and high hop bitterness, with a bit of diacetyl (a naturally occurring byproduct with a buttery aroma and flavor) and low caramel flavor.

- ✔ **Commercial examples:** Sierra Nevada Pale Ale, California; Great Lakes Burning River Pale Ale, Ohio.

Beginners, try making this beer by using two 3.3-pound cans of pale, unhopped extract. You can add hop character by boiling 2 ounces of Northern Brewer hops for 1 hour and 1 ounce of Cascade hops in the last 5 minutes of the boil for the classic American pale ale character. When you're confident in your brewing abilities, make the American pale ale recipe later in this chapter.

Brown porter

The porter name is borrowed from a group of people known to consume large quantities of this beer: the porters at London's Victoria Station. The porters had a habit of ordering portions of several beers mixed into the same drinking glass. This concoction came to be known as *entire.* One enterprising brewer capitalized on this habit by marketing a beer that closely approximated this blend of brews, and he used the name *porter* to identify it.

- ✔ **Appearance and flavor profile:** Medium to dark brown. Fruity esters are acceptable, and hop flavor and aroma are nonexistent to medium. No roasted barley or strong burnt character is expected. Low to medium malt sweetness, with medium hop bitterness and low diacetyl. Light- to medium-bodied.

- ✔ **Commercial examples:** Samuel Smith Taddy Porter, England; Young's London Porter, England.

When you first try your hand at porter brewing, try producing the color and flavor by adding specialty grains (dark caramel, chocolate malt, black malt, or roasted barley) to your pale malt extract rather than just buying a dark malt extract — the result is more satisfying. When you're ready to take the next step, try making a brown porter (see the recipe later in this chapter for details).

Stout

Stout is a hearty, top-fermented beer strongly associated with the British Isles and Ireland; it's known for its opaque-black appearance and roasty flavors.

- ✔ **Appearance and flavor profile:** Opaque black. Medium-bodied, with no hop flavor or aroma. Roasted barley character is expected. Slight malt sweetness or a caramel malt character is okay. Medium to high hop bitterness with a slight acidity or sourness is possible. A very low diacetyl level is okay.

- ✔ **Commercial examples:** Guinness Stout, Ireland; Murphy's Stout, Ireland.

The first time you attempt to brew a stout, try producing the color and flavor by using specialty grains (dark caramel, chocolate malt, black malt, or roasted barley) rather than buying a dark malt extract — the result is more satisfying. Later on, we recommend brewing the stout recipe at the end of this chapter.

English India pale ale

One particular substyle of English-style pale ale is known as *India pale ale,* or IPA for short. The beer gets its name from Britain's colonial presence in India during the 1800s.

- ✔ **Appearance and flavor profile:** Golden to light copper. Very fruity and estery. Medium maltiness with high hop bitterness and low diacetyl. Hop flavor and aroma are medium to high. Medium-bodied, with evident alcoholic strength.

- ✔ **Commercial examples:** Samuel Smith's India Ale, England; Fuller's IPA, England.

Hoppiness is the hallmark of this brew. Most beer enthusiasts say that the hoppy flavor leaves other brews lacking once you hit the perfect mark. English IPA is a bit less delightfully bitter than the American IPAs, so choose accordingly.

American premium lager

The pale American-style premium lager is the most-produced beer style in the United States. Despite the categorical name, these beers are rather one-dimensional compared to those made in Europe and elsewhere. The reasons for this condition are largely the cheaper ingredients used to make them and the treatment of beer in America as a beverage designed for mass consumption. American craft brewers, on the other hand, are producing premium lagers that are more deserving of the name.

- ✔ **Appearance and flavor profile:** Very pale to golden. No fruitiness, esters, or diacetyl. Low malt aroma and flavor are okay. Low hop flavor or aroma is okay, but low to medium bitterness is expected. Effervescent and light-bodied.

- ✔ **Commercial examples:** Leinenkugel's, Wisconsin; Brooklyn Lager, New York.

Beginners, start with 5 pounds of the palest extract you can find and add 0.5 pound of brewer's rice syrup. Use high-quality lager yeast (preferably liquid) and allow a long, cool fermentation and aging period. When you're ready to try intermediate brewing techniques, we recommend the recipe later in this chapter.

Märzen/Oktoberfest

Oktoberfest beer, as an individual style, is an offshoot of another, larger lager style known as *Märzen* or *Märzenbier*. This fairly heavy, malty style is brewed in the spring and named for the month of March (März). It was often the last batch of beer brewed before the warm summer months. This higher-gravity beer was then stored in Alpine caves and consumed throughout the summer. Whatever beer was left in storage at harvest time was hauled out and joyously consumed.

- ✔ **Appearance and flavor profile:** Amber to coppery orange. No fruitiness, esters, or diacetyl are evident. Low hop flavor and aroma are okay. The malty sweetness boasts of a toasty malt aroma and flavor. Medium-bodied, with low to medium bitterness to keep the malty character from becoming cloying.

- ✔ **Commercial examples:** Wurzburger Oktoberfest, Germany; Capital Oktoberfest, Wisconsin.

Oktoberfest/Märzenbiers are malt-accented beers. When you're just beginning, start with at least 6 pounds of pale extract and steep 1.5 pounds of 20-L crystal malt and 0.5 pound toasted malt in water and then strain into wort. Use high-quality lager yeast (preferably liquid) and allow a long, cool fermentation. Later, give the Oktoberfest recipe a try.

Traditional bock

Traditional bock beer is a hearty, bottom-fermented beer with a generously malty character and burnt-toffee, dark-grain flavors. It has a creamy mouth feel, and the finish is lengthy and malty sweet. Hop bitterness is subdued — it's just enough to cut the malt's cloying character. The color can run the spectrum from a deep burnt orange to mahogany. The alcohol content is usually considerable; a true German bock beer must have a minimum alcohol content of 6.5 percent to be called a bock.

- ✔ **Appearance and flavor profile:** Deep copper to dark brown. No hop aroma, fruitiness, or esters are evident. The malty-sweet character predominates in aroma and flavor, with some toasted chocolate-malt character. Low bitterness, low hop flavor, and low diacetyl are okay. Medium- to full-bodied.

- ✔ **Commercial examples:** Spaten Bock, Germany; Einbecker Ur-Bock, Germany.

Traditional bock beers are big beers; we suggest beginners start with no fewer than 8 pounds of amber extract, steep 1.5 pounds of 40-L crystal, and strain into wort. Use high-quality lager yeast (preferably liquid) and allow a long, cool fermentation period. If you love traditional bock beers, check out the recipe later in this chapter.

Herb, spice, and vegetable beer

Although you don't see many herb and spice beers on your local beer retailers' shelves, a few do exist. If homebrewers had their say, many more would be available. The herb-and-spice-beer category is one of the more popular among homebrewers because it presents an almost unlimited number of choices.

Herb and spice beers may include lemon grass, ginger, cumin, allspice, caraway, mace, pepper, cinnamon, nutmeg, or clove, among myriad other additions. Vegetables in beer, on the other hand, are very few and far between — and, to my knowledge, no such thing as a vegetable extract is made for brewing — so you're pretty much limited to pumpkin and hot peppers here.

- ✔ **Appearance and flavor profile:** Herb, spice, and vegetable beers are often made with an anything-goes approach. In light of this practice, we have no way to accurately describe what to expect from one of these beers.

- ✔ **Commercial examples:** Left Hand Good Juju Ginger, Colorado; Fraoch Heather Ale, Scotland.

Here's an opportunity to go a little crazy. If you're a beginning brewer, pick a favorite herb or spice and add it to your beer either in the brewpot (in the last 20 minutes of the boil) or during the fermentation and aging phase. *Hint:* Too little is better than too much; you can always add more to the next batch you make. Add your veggies in the secondary fermenter for best results. More advanced brewers should try the spice beer recipe later in this chapter.

Christmas/winter/spiced beer

Many breweries produce unique seasonal offerings that may be darker, stronger, spiced, or otherwise fuller in character than their normal beers. The special ingredients should complement the base beer and not overwhelm it.

- ✔ **Appearance and flavor profile:** These winter seasonal brews are often made with an anything-goes approach, so accurately describing what to expect from one of them is difficult.

- ✔ **Commercial examples:** Anchor Our Special Ale, California; Harpoon Winter Warmer, Massachusetts.

Yule ales and winter warmers are made to toast the holidays and the winter season. Warm spice flavors and elevated alcohol levels are pretty effective at putting a flush in your cheeks. Beginners, start with at least 7 pounds of pale malt extract and then add whatever adjunct grains or flavorings evoke the holiday spirit for you. When you've had more practice, try the recipe later in this chapter.

Trying a Few Beer Recipes

What would a good book on homebrewing be without the recipes? Following are some basic recipes for the homebrew enthusiast. Your favorite flavor profiles are listed, with plenty of room for experimentation. Enjoy!

Irish Red Ale

Malt extract:	6.6 pounds Northwestern Gold extract
Specialty grain:	1 pound 60-L crystal malt
	⅛ pound roast malt
Bittering hops:	0.5 ounce Fuggles (60 min)
	0.5 ounce Fuggles (40 min)
Flavoring hops:	1 ounce Fuggles (20 min)
Finishing hops:	0.5 ounce Kent Goldings (5 min)
Yeast:	Wyeast #1084
Primary:	6 days at 65°
Secondary:	12 days at 65°

Brewer: Marty Nachel

American Pale Ale

Malt extract:	6 pounds Northwestern light
Specialty grain:	2 pounds 40-L crystal malt
Bittering hops:	1 ounce Northern Brewer (60 min)
Flavoring hops:	1 ounce Northern Brewer (30 min)
	1 ounce Spalt (15 min)
Finishing hops:	1 ounce Cascade (5 min)
Dry hop:	1 ounce Cascade
Yeast:	Wyeast #2112
Primary:	7 days at 65°
Secondary:	21 days at 65°

Brewer: Marty Nachel

Brown Porter

Malt extract:	3.3 pounds Brewmaker Mild hopped extract
	3.3 pounds Munton and Fison amber
Specialty grain:	1.66 pounds caramel malt
	5 ounces chocolate malt
Bittering hops:	0.5 ounce Northern Brewer (8 AAU) (35 min)
Finishing hops:	0.25 ounce Northern Brewer (8 AAU) (2 min)
Dry hop:	0.25 ounce Cascade
	0.5 ounce Hallertauer
Yeast:	Wyeast #1028
Miscellaneous fermentable ingredients:	6 ounces barley syrup
Water treatment:	0.5 teaspoon noniodized salt
Primary:	12 days at 65°
Secondary:	(not given)

Brewer: Dennis Kinvig

Award won: 1st Place, AHA Nationals

Stout

Malt extract:	6.6 pounds amber extract
Specialty grain:	0.5 pound black malt
	0.5 pound 40-L crystal malt
	0.25 pound roasted barley
Bittering hops:	2 ounces Fuggles (60 min)
Finishing hops:	1 ounce Willamette (10 min)
Yeast:	Wyeast #1084
Primary:	(not given)
Secondary:	(not given)

Brewer: Northwestern Extract Co.

English India Pale Ale

Malt extract:	6.6 pounds Northwestern extract
Specialty grain:	1 pound 40-L crystal malt
	⅛ pound roast malt
Bittering hops:	1.5 ounces Northern Brewers (60 min)
Finishing hops:	1 ounce Kent Goldings (10 min)
Dry hop:	1 ounce Fuggles
Yeast:	Wyeast #1098
Miscellaneous flavoring ingredients:	8 ounces maltodextrin powder
Primary:	1 week at 65°
Secondary:	2 weeks at 65°

Brewer: Marty Nachel

Typical/unusual procedures used: Add 1 ounce Fuggles hops and maltodextrin powder to secondary fermenter.

American Premium Lager

Malt extract:	5 pounds Alexander's pale extract
Specialty grain:	1 pound 10-L crystal malt
Bittering hops:	1 ounce Northern Brewer (60 min)
Flavoring hops:	1 ounce Perle (20 min)
Finishing hops:	0.5 ounce Saaz (5 min)
Yeast:	Wyeast #2035
Miscellaneous fermentable ingredients:	1 pound brewer's rice syrup
Fining agent/clarifier:	2 teaspoons Irish moss
Primary:	6 days at 60°

Brewer: Marty Nachel

Oktoberfest

Malt extract:	6.6 pounds Bierkeller extract
	1 pound amber DME
Specialty grain:	0.5 pound 10-L crystal malt
	0.5 cup chocolate malt
Bittering hops:	1 ounce Cascade (60 min)
Flavoring hops:	1 ounce Hallertauer (30 min)
Finishing hops:	0.75 ounce Tettnanger (1 min)
Yeast:	Wyeast #2206
Primary:	11 days at 50°
Secondary:	10 days at 45°
Tertiary:	15 days at 35°

Brewer: John Janowiak

Award won: 1st Place, AHA Nationals

Traditional Bock

Malt extract:	6 pounds pale liquid malt extract
	2 pounds pale DME
Specialty grain:	1 pound 60-L crystal malt
	0.5 pound chocolate malt
Bittering hops:	0.5 ounce Hallertauer (60 min)
	0.5 ounce Hallertauer (45 min)
Flavoring hops:	0.5 ounce Hallertauer (30 min)
Yeast:	Wyeast #2206
Fining agent/clarifier:	1 teaspoon Irish moss
Primary:	8 days at 55°
Secondary:	14 days at 45°

Brewer: Marty Nachel

Herb Spice and Vegetable Beer

Malt extract:	10 pounds Northwestern light
Specialty grain:	1 pound 40-L crystal malt
Bittering hops:	2 ounces Mount Hood (60 min)
Flavoring hops:	1 ounce Northern Brewer (30 min)
Finishing hops:	1 ounce Cascade (5 min)
Dry hop:	1 ounce Cascade
Yeast:	Pasteur Champagne (dry)
Miscellaneous flavoring ingredients:	2 sticks cinnamon, 1 teaspoon cloves
Primary:	11 days at 60°
Secondary:	25 days at 60°

Brewer: Marty Nachel

Award won: 1st Place, Dukes of Ale Spring Thing

Christmas/Winter Spiced Beer

Malt extract:	3.5 pounds Munton & Fison stout kit
	3.3 pounds Munton & Fison amber extract
	3 pounds Munton & Fison amber DME
Bittering hops:	0.5 ounce Hallertauer (55 min)
Finishing hops:	0.5 ounce Hallertauer (5 min)
Yeast:	Wyeast #1007
Miscellaneous fermentable ingredients:	0.75 pound honey
Miscellaneous flavoring ingredients:	Five 3-inch cinnamon sticks
	2 teaspoons allspice
	1 teaspoon cloves
	6 ounces grated ginger root
	6 medium oranges (rinds only)
Primary:	14 days at 60°
Secondary:	(not given)

Brewer: Philip Fleming

Award won: 1st Place, AHA Nationals

Typical/unusual procedures used: Simmer all the flavoring ingredients in the honey for 45 minutes; strain into the brewpot.

Part VI
The Part of Tens

the
part of
tens

In this part . . .

✔ Get the scoop on what to do when something goes wrong with your fermentation.

✔ Understand how fermented foods can improve your immunity, digestion, and overall vitality, and even save you money.

✔ Discover the various resources for getting started with your ferment.

✔ Find out ways to live a long and healthy life.

Chapter 18

More Than Ten Troubleshooting Tips for Fermented Creations

In This Chapter

▶ Making sure your ferment turns out just right

▶ Avoiding messes and disasters

*V*ery few things in life go perfectly as planned. And fermenting is one of those endeavors that requires patience and sometimes a little trial and error. When things don't seem to be working as you expected, check out the tips in this chapter.

Special thanks to Jenna Empey and Alex Currie, owners of Pyramid Farm & Ferments in Prince Edward County, Ontario, Canada, for contributing some really helpful information. Pyramid Farm & Ferments practices the traditional art of fermentation to create handcrafted, local, raw, and cultured foods. Go with your gut! Check out their website at www.pyramidfarmandferments. com if you'd like to learn more.

My Fermented Food Is Too Salty. What Do I Do?

Upon tasting your fermented food, if you find it extremely salty, try rinsing it in a little water before eating. Taste as you go to find the right salt level for you.

Next time you ferment, taste your way through the salting process to see what works for you. Salt is necessary for lacto-fermentation because it encourages an environment where the lactobacillus can thrive and the vegetables will ferment properly. But too much salt is undesirable.

A rough guideline is 3 tablespoons of salt to 5 pounds of shredded cabbage.

Why Is the Fermentation Taking So Long?

There's no set time of completion for any fermented creation. Temperature, humidity, ingredients, and your environment all affect the fermentation process.

Some fermented foods take a few days and others a few months to reach the right flavor, but remember that time is an essential ingredient in fermentation.

Taste your fermented creation along the way and experience its different stages of development. Also know that what is ready for some may not be ready or sour enough for you. By tasting it through the fermentation process, you get a better idea of how long different types of fermented foods take and what level of sour you enjoy.

Why Are My Fermented Creations Different throughout the Year?

Warmer temperatures accelerate the fermentation process, and cooler temperatures slow it down. In the summer months, find a cooler place for your ferment vessel and, when it's finished fermenting, refrigerate it. Refrigeration helps slow down the fermentation process, stabilize the ferment, and hold the desired flavor.

In the winter months, fermentation slows down and things take longer to sour. Find a warmer place to keep your fermenting vessel. Aim for an environment that's approximately 65 degrees or higher all year round.

Why Is My Ferment Too Soft or Mushy?

A few factors can contribute to your fermented food being too soft or mushy:

- ✔ It fermented too fast because the temperature was too high.
- ✔ You didn't add enough salt.

Soft or mushy fermented food isn't necessarily spoiled. Some people prefer a softer ferment, but if the texture puts you off, try cooking with it.

Why Isn't My Ferment Working?

A few factors can derail your fermentation:

- ✔ Did you give it enough time? Some fermented creations take weeks to months to taste just right.

- ✔ Is it too cold in your fermentation location? Anything below 50 degrees will be very slow to get going.

- ✔ Did you use iodized salt? Iodine has antibacterial properties and has been known to affect fermentation. Use a salt that's free of running agents and iodine for best results.

- ✔ Is your tap water chlorinated? If you're making a ferment that involves a brine, use nonchlorinated water.

- ✔ How did you sterilize your equipment? It's important to have clean fermentation vessels and equipment, but avoid using harsh or antibacterial sterilization solutions. Steam and boiling water do a fine job.

Why Is My Fermented Creation Too Dry?

Often your fermented creation can start out very juicy and full of brine, only to mysteriously dry up a few days in. As the salt draws out the juice from the vegetables, the brine increases for the first three days of fermentation. This is usually followed by a recession of the brine.

This is where weighting your ferment comes in. Weights are very important; they help keep the brine up and the vegetables submerged beneath the brine, which is essential for proper fermentation. In smaller jars you can use a sealable plastic bag filled with water that sits on the top level of the fermenting food, acting as a weight. You can also use boiled rocks, exercise weights, or other jars to help keep the vegetables submerged.

What Do I Do about Yeast or Mold on the Surface of the Ferment?

If you see bacterial growth on the surface of your ferment, don't panic. This is a common thing. People often overreact to any presence of mold or yeast growth, but such growth doesn't necessarily mean your ferment is ruined.

If you see a white film, foam, or chunks of a mold bloom, get a spoon and scoop it out the best you can. If it comes back in a few days, repeat.

If the mold has occurred in a larger way and the ferment has dried out or looks discolored on the top, remove a few inches off the top layer and see whether lower layers look better. Most often, a few inches below the affected area you have a perfectly good and delicious fermented food!

Make sure you keep your ferment covered to prevent insects and other contaminants from getting into it. A piece of cloth does just fine.

What Should I Do about a Ferment Jar That's Bulging?

If you're fermenting in a sealed vessel or jar, it's common for it to bulge or leak. This is caused by the buildup of carbon dioxide as the fermentation occurs. Too much pressure building up can be a dangerous thing if it causes the jar to swell and break. You don't have to use a sealed vessel for fermentation. A cloth-covered, weighted vessel works well.

If you notice your jar bulging, open it over the sink and release the pressure. It may bubble or leak. Take a fork and press the vegetables down under the brine again.

Why Did the Color Change?

The appearance of your fermented creation at the beginning of its fermentation time as compared to the end result can often be completely different!

Pink or red vegetables will turn your ferment into a variety of shades of pink, red, and purple. Some green vegetables naturally brown a bit as they ferment. Colors can fade or intensify based on the ingredients you choose.

If your batch of ferment has all turned brown, smells off, or tastes bitter, this is a sign that it has spoiled, and it should be composted.

Why Is My Ferment Leaking or Overflowing?

As the salt draws the liquid out of the shredded vegetables, the ferment's moisture content increases, causing your brine to rise and spill over your vessel. Generally this peaks three days after you pack the ferment into your fermentation vessel. Other factors, such as closeness to the sea and a full moon, may affect moisture content. After the brine rises it often recedes, which can dry out your ferment. Use a fork to press the vegetables back under the brine for proper fermentation.

Always leave a few inches of space at the top for the expansion, and place a plate or bowl under your vessel to catch any leaks.

Why Does It Stink?

Fermentation is a process full of noticeable, changeable, and intense aromas, but don't mistake a strong odor for a bad odor! If the smell is taking over your house, try keeping your fermenting vessel in a different and lower traffic area.

If it's spoiled, the smell will be so bad that you'd never dream of eating it.

Chapter 19

Top Ten Benefits of Eating Fermented Foods

. . . .

In This Chapter

▶ Improving your health by eating fermented foods

▶ Saving time and money by fermenting

▶ Making a positive environmental impact

. . . .

*I*f you've read through any portion of this book, you know that we think fermented foods are awesome. In this chapter, we present ten reasons why you should at least consider getting started with fermented foods. For more details on the health benefits of fermented foods, flip to Chapter 3.

A Much-Needed Nutritional Boost

Fermented foods are nutritionally active and offer your body a mega-dose of vitamins and nutrients. The process of lacto-fermentation creates new nutrients in foods, increases the bioavailability of minerals and nutrients, and increases vitamin absorption.

Digestion, Enzymes, and Probiotics

Lacto-ferments improve your digestion by delivering beneficial lactobacilli to your digestive system. Because fermented foods are a way of preserving food while keeping the lactobacilli active, the enzymes that are created are extremely beneficial to your health. Lacto-ferments also help normalize your stomach's acidity and stimulate the production of beneficial intestinal flora. This makes it easier for your body to break down and assimilate the foods you eat, fermented or not.

Immunity Boost

Eating lacto-ferments stimulates and improves your digestive system. Eighty percent of your immune system is in your digestive system, so when you have a healthy gut, you have a healthy immune system, which means that you'll be healthy.

Unique Flavor

Fermenting adds a tangy, unique flavor to foods, and fermented foods complement and add flavor to any meal. You always want to stimulate and encourage your palate with new flavors, and fermented foods can do just that.

Money Savings

By fermenting your own food, you save on the amount of times that you need to buy fresh produce. For instance, one head of cabbage yields a good amount of sauerkraut, and it lasts for months.

Time Savings

Fermented items do take some time to prepare, but after they're ready, many of them last a long time. This saves you from preparing all your meals from scratch.

Ecological Impact

By fermenting, you don't make as many trips to the grocery store, which means you use less packaging, save energy and resources, and don't rely on foods to be shipped. For the most part, you're embracing local and organic sources of foods, which only has a positive impact on the environment.

Slow Food Movement

Fermenting your own foods definitely embraces the slow food movement. Not only are the ingredients encouraged to come from local farmers or other organic sources, but fermented foods are also anything but fast foods. Because fermentation doesn't require the chemicals, processing, or high-heat temperatures that some fast cooking methods require, fermentation embraces time and getting the final result just right. This process is considered slow, which is a good thing!

Control over Your Food

Taking the time to connect with your food and be involved in the process from start to finish is a necessary step in the preservation of your health. When you know what goes into your food, you know what's going into your body.

The Satisfaction of Doing Something Good for Yourself

You definitely get a sense of gratification from fermenting your own foods. You also get a sense of satisfaction from improving your health, whether it be your digestion, energy, immunity, or overall vitality. So try a recipe or two and start fermenting today!

Chapter 20

More Than Ten Food and Equipment Resources

· ·

In This Chapter

▶ Getting fermenting equipment and supplies online

▶ Tapping resources to brew beer, make wine, and make sausage

· ·

*T*hroughout this book, we've tried to make fermenting a fun and easy process. We think everyone who loves food or is concerned about their health should give fermenting a try. Unfortunately, acquiring the ingredients and equipment can be somewhat of a chore. That's where this chapter comes in. The resources listed here should provide all the help you need.

Cultures for Health

This comprehensive online resource has all the information you need to know about fermenting, with many of the tools and supplies available for order online. It also has many videos and resources for added info to make your fermenting experience effortless.

807 N. Helen Avenue
Sioux Falls, SD 57104
Phone: 800-962-1959
Website: www.culturesforhealth.com

Yolife

This company sells yogurt starters and yogurt makers for both dairy and nondairy cultures.

Tribest Corporation
P.O. Box 4089
Cerritos, CA 90703
Phone: 888-254-7336
Website: www.yolifeyogurt.com/index.asp

Water Kefir Grains

Check out this website for your first starter kit to make water kefir. The kit comes with the grains all ready to go and instructions on how to get started.

Website: www.waterkefirgrains.com

Wildwood Foods

This is a fabulous company for organic soy products. Tempeh, tofu, and other items are available. You can find Wildwood products at many super-markets and health food stores across North America.

Phone: 800-588-7782
Website: www.wildwoodfoods.com

Miso Master

This company sells one of the most delicious miso pastes available. You can get an assortment of flavors and varieties, and for those who are sensitive to soy, the company even offers a chickpea-based miso. You can find many of this company's products at superstores and local health food markets across the country.

Great Eastern Sun
92 McIntosh Road
Asheville, NC 28806
Phone: 800-334-5809
Website: www.great-eastern-sun.com

Leeners

This is a one-stop-shop for all things brewing and fermenting. You can supply your kitchen with all the tools of the trade, including the fun and funky ingredients that make your hobby enjoyable, with a bunch of great recipes included for free.

Leeners
9293 Olde Eight Road
Northfield, OH 44067
Phone: 800-543-3697
Website: www.leeners.com

Homesteader's Supply

This company carries a huge assortment of equipment for fermenting, cheesemaking, canning, and brewing. This site has a little bit of everything you need, including kits and cultures. The staff is very knowledgeable and the prices are reasonable.

Homesteader's Supply
P.O. Box 5369
Chino Valley, AZ 86323
Phone: 928-583-0254
Website: www.homesteadersupply.com

The Sausage Maker

Looking for supplies can be the most difficult part of sausage making. This site covers everything you need, things that are both helpful and fun! Check out the prices on kits and hardware to take your sausage-making hobby to a whole new level.

The Sausage Maker
1500 Clinton Street, Building 123
Buffalo, NY 14206-3099
Phone: 888-490-8525
Website: www.sausagemaker.com

New England Cheesemaking Supply

This is the premier site for anything you could possibly need to make cheeses. The supplies and reference material are enough to keep you coming back. The cultures and periodicals that are aimed at improving your technique and knowledge of cheeses make this the best online site for anything to do with home cheesemaking.

New England Cheesemaking Supply Company
54B Whately Road
South Deerfield, MA 01373
Phone: 413-397-2012
Website: www.cheesemaking.com

The Sausage Source

This is a great site for sausage-making kits. The kits are complete and include the tools as well as the ingredients. Buy a kit and get started right out of the box — just add your meat.

The Sausage Source
3 Henniker Road
Hillsboro, NH 03244
Phone: 800-978-5465
Website: www.sausagesource.com

Adventures in Homebrewing

No matter whether you're looking to brew beer or make wine at home, this site has everything you could possibly need. It offers reasonably priced kits for beginning to expert homebrewers, as well as winemaking kits and supplies.

Adventures in Homebrewing
23869 Van Born Road
Taylor, MI 48180
Phone: 313-277-2739
Website: www.homebrewing.org

Chapter 21

More Than Ten Tips for a Long and Healthy Life

As you've seen throughout this book, fermented foods can improve your health. But fermented foods are just one component of living a healthy life. This chapter gives you some tips for making the most of your fermented food and discusses other healthy habits you may want to embrace.

Food Is Medicine, So Eat to Enhance Your Health

Fermented foods are great for your gut. We're big believers in preventive healthcare and food as medicine. *Preventive healthcare* means taking small, everyday actions to maintain your overall health to avoid major issues with disease in the future and to protect yourself from infection and stress, or at least minimize your encounter with it. This starts with choosing great foods for your body. It's true what they say — you are what you eat — so why not choose health? Fermented foods can play a big role in your personalized preventive healthcare plan and everyday digestibility.

Use Alternative Sugar

Did you know that alternative sugars can help sustain your energy longer? When using any sugar for your home recipes, choose *unrefined sugars* such as raw honey, maple syrup, or Sucanat (made of whole cane sugar). Honey is a natural antibacterial and often has other properties that can cause interference with fermentation, so be open to experimentation or dilute it where possible. Choosing unrefined sugar ensures that your foods come packed with the beneficial vitamins and minerals you need rather than the empty calories from processed sugars.

Reduce Plastic Use and Go BPA-Free

Plastic has given the modern world tons of benefits but has also been the cause of many environmental and health concerns. Be sure to get *food-grade safe plastic* containers when making or storing any food item. Try to reduce your use of plastic in your life and rid yourself of any leaching chemicals that may enter your body. Go BPA-free when possible, because bisphenol A has been linked to many health-related problems. Find food and drinking containers that are made from glass or lined with stainless steel.

Choose Organic

Fruit and vegetables may require more or less pesticides depending on their growing conditions and environmental needs. Today, many farmers rely on these, yet plenty of studies have shown that pesticides have harmful effects on the planet and your health.

It's always recommended that you wash your produce well before eating it, but you can go one step further and try to increase the amount of organic produce you buy. Apples, strawberries, celery, spinach, and peaches are among the top-rated for being heavily sprayed. Do your research on various organic certifications and try to integrate even just a few organic foods into your diet. You'll taste the difference and feel great for making one small step in a positive direction.

Get to Know Your Farmer

Have you ever picked and tasted fruit fresh from the tree or picked spinach straight from the soil? It sure is different from the experience of buying it in the grocery store. Food is your main source of fuel, yet few people see it through its entire life cycle.

From seed to plate, your food often travels a long way and through a long process before you get a chance to eat it. Take a moment and think of all the people and processes involved in just one food item: Sowing seeds, selling seeds, weeding, harvesting, watering and maintaining, processing, distributing, supplying, and so on.

You should be grateful for farmers worldwide and treat them with the utmost respect. The demand for local and fair-trade food is on the rise, and with it many different farming certifications. Educate yourself on the various certifications or plan a trip to your local organic and nonorganic farmer. Every farmer has different practices, values, and methods. The more you know about the food and where it's from, the more you can make informed decisions on what goes into your body. Be daring and ask questions!

Be Conscious about Your Condiments

Have you ever made your own chutney, salsa, or condiments before? You're sure to notice the fresh flavor difference. The condiments that you purchase off the shelves of grocery stores don't always have the same love and care you can give them at home. They're made to last long and packaged for preservation, but this can come with a compromise of ingredients or high-quality taste. The shelf life may be long, but preservatives are probably at play. You don't always know where the produce in your condiments even came from. Plus, condiments often come in plastic packaging that isn't biodegradable or takes a long time to recycle.

The beauty of homemade fermented condiments is that you can control the preservatives and the high sugar content as compared to store-bought ones. They may not always last as long, but they surely give you great nutrients, and you can package them in glass jars that you can reuse in future canning.

Eat Whole Grains

Around the world, each meal is filled with an added grain such as rice, wheat, oats, quinoa, or rye. How many of these grains are whole grains? Grains are a major part of the world's diet, so you should ensure that you're getting the utmost nutrients from them. With whole grains, you get the essential nutrients of the entire grain seed. If the grain has been processed, crushed, or cooked, it should give the same benefits if everything is still in the food. Look for whole grains that haven't been processed to get the best balance and nutrients from each grain seed.

Choose GMO-Free Foods

Do you know exactly how your food was made? Do you want the best health benefits from your recipes? Certified GMO-free produce helps you avoid any unwanted changes that happened to your food before it was placed on the shelf.

Sometimes, food is changed to enhance its shelf life, freezing point, color, texture, or other factors that make it more marketable. Some of these changes are still being tested, and the long-term effects aren't well known. Do some research and decide for yourself whether GMO-free food is the right choice for you. When buying produce for your homemade recipes, we suggest that you do your best to choose GMO-free for the purest food choices for your body's needs.

Get Protein from Plants

One of the best sources of protein can come from plants. More specifically, nuts and seeds are a great protein alternative. They provide healthy fats and proteins that can help lower body fat but fill the body with good cholesterol. We suggest you soak your nuts and seeds to increase their digestibility and boost your nutrient intake. Nuts and seeds are great as a snack on the go or sprinkled on your morning cereal. Go nuts!

Find Vegan Milk, Butter, and Dairy Options

If you want to change milk in a recipe for a nondairy version, opt for rice milk, almond milk, coconut milk, or even hemp milk. Milk can also be made from oats and quinoa. Many of these milks can be store-bought or homemade. Coconut oil is the perfect replacement for butter. It's rich, creamy, and has a fabulous texture. You can replace coconut oil cup for cup with butter for your vegan-friendly recipes. Coconut oil is a plant-based saturated fat that's a natural, nourishing alternative to more processed vegetable oils.

Learn to Love Water

Water may seem flavorless, but it's essential to your well-being and can certainly quench your thirst. Want to sweeten it up a bit? Try adding a squeezed lemon in your water with a spoonful of maple syrup. You'll learn to love every last drop.

Index

• *S* •

About the Authors

Certified chef and culinary nutritionist, **Marni Wasserman** uses passion and experience to educate individuals on how to adopt a realistic, plant-based diet that is both simple and delicious. She is dedicated to providing individuals with balanced lifestyle choices through organic, fresh, whole, and natural plant-based foods.

As a prominent figure of health and nutrition in Toronto, Marni is a contributor to *Chatelaine, Huffington Post, Bamboo,* and *Tonic Magazine.* She has made several TV appearances on CBC and CTV, and has been featured in the *Toronto Star.* Marni has also consulted with the Windsor Arms Hotel for its vegan and vegetarian menus, and she participates in several live speaking and cooking demonstrations. If that isn't enough, she is also the author of several well-received plant-based e-books such as *Cleansing with Super Foods* and *Veggin' Comfortably.*

Marni is a graduate of the Institute of Holistic Nutrition in Toronto and the Natural Gourmet Culinary School in New York.

Marni is also the owner of her very own flagship Food Studio and Lifestyle Shop in Toronto. This is where Marni teaches her signature plant-based cooking classes, collaborative workshops, and retreats. Her studio is also a place where people can come to support their lifestyle with Marni's delicious knowledge, sustainable living eco-inspired products, and of course, nourishing food.

You can learn more about Marni by visiting her on Facebook, Twitter, or www.marniwasserman.com.

Amy Jeanroy is a freelance writer and newspaper editor who has been making and eating fermented food for 20 years. She is passionate about filling the pantry with homemade, healthy foods. Amy is the owner and managing editor of www.thefarmingwife.com, where she shares her daily recipes for homemade goodness from the kitchen.

Amy is a Master Gardener, food writer for Foodista.com, and an herb garden writer for About.com.

Dedication

Marni Wasserman: I would like to dedicate this book to all the people that have contributed to the foundation of my knowledge in whole foods including all my instructors at the Institute of Holistic Nutrition, especially Patricia Meyer Watt, one of the first people to introduce me to fermented foods.

My parents, family, and friends have always been supportive of my lifestyle and interest in health-focused foods. My interest in fermented foods is just one more piece of the puzzle that adds to my intricate web of knowledge. Over the years, my interest in fermentation has also developed as a result of reading Sandor Katz and Sally Fallon. Their traditional roots on whole foods have been a huge contribution to my interest in the subject.

Amy Jeanroy: This book is dedicated to my husband Cal, whose support and encouragement have been unwavering. A thanks also goes out to Nathanial, Gabriel, Josiah, Sebastian, and Rebekah, who usually fall asleep to the sound of my keyboard. I could not do this without my wonderful family's understanding.

Authors' Acknowledgments

Marni Wasserman: Robin Newman and Sandra Braun were extremely helpful with their organization and creativity, which helped get this book on track and keep it on time. These girls were key players in making sure everything came together! With their unique editing and writing styles and attention to detail, they made this project fun and approachable.

A special thanks to Tracy Boggier, who saw the opportunity for me to be the plant-based contributor to *Fermenting For Dummies*.

Amy Jeanroy: This book was a labor of love, but it was also a wonderful collaboration with fellow author, Marni Wasserman. Her energy kept me inspired throughout the project. Thanks to my agent Barb Doyan, who always knows the perfect book topics for me, and of course to John Wiley & Sons, for offering such a wonderful platform for us writers.

Publisher's Acknowledgments

Acquisitions Editor: Tracy Boggier

Senior Project Editor: Tim Gallan

Copy Editor: Todd Lothery

Technical Editors: Pam Mitchell, Mike Tully, Emily Nolan

Art Coordinator: Alicia B. South

Photographer: T. J. Hine

Project Coordinator: Kristie Rees

Cover Image: ©T. J. Hine Photography